Liang Shaoji's Silkworm Art

This book focuses on the work of Chinese contemporary artist Liang Shaoji and emphasises the contribution of multispecies ethnography to art criticism. Over three decades, Liang has worked with domestic silkworms to craft art that embodies the Daoism-inspired ecological motif of 'ziran'. Are silkworms co-authors or alienated fabricators in such creative practice? Based on a multi-sited ethnographic study conducted in China, the book delves into Liang's artistic techniques involving close collaboration with silkworm farmers and biologists. In doing so, it makes a significant contribution to discussions of non-human agency and labour. The author unveils the intricate power dynamics between silkworms and their caretakers, revealing multi-sensory knowledge, anthropomorphic kinship, and moral dilemmas inherent in working with these insects. This volume will be of particular interest to scholars working in the anthropology of art, human-animal studies, and environmental humanities.

Feixuan Xu is an anthropologist specialising in the anthropology of art and multispecies ethnography. She obtained a PhD in Creative Media from City University of Hong Kong and was Asymmetry Postdoctoral Research Fellow at The Courtauld Institute of Art, UK.

Multispecies Anthropology: New Ethnographies
Series Editors: Rebecca Cassidy and Garry Marvin

Human-Animal Studies has come of age. Intrinsically inter or transdisciplinary, it has encouraged scholars from diverse disciplines to use a range of approaches and resources to explore the relationships that humans have with other animals and forms of life and how these are experienced and expressed.

Animals have traditionally appeared in anthropology as aspects of cosmological systems, essential to livelihoods including hunting, fishing, pastoralism, herding and agriculture; significant in economic systems as wealth and for exchange; celebrated in sports and other forms of entertainment and so on. More recently, animals and other forms of life have been brought to the foreground, where they are framed as important in and of themselves, rather than as reflections of other, more important relationships between people. This series is interested in this approach and the wider questions it poses about subjectivity, representation and anthropological theory.

Uniquely, this series focuses exclusively on monographs which are based on first-hand, sustained, ethnographic fieldwork: an approach which allows for the exploration of the intricacies and immediacies of lives with other animals. It brings together detailed accounts of how humans experience, engage with, live with, other animals, but also with plants and other living matter, generated within particular social and cultural worlds as they are captured by fieldwork. This format will enable authors to pursue some of the most important questions about our lives with others – and the role that anthropology might play in our futures together.

The Multispecies Triad of Cattle Ranching
A Horseback Ethnography
Andrea Petitt

Liang Shaoji's Silkworm Art
Untangling Multispecies Craft
Feixuan Xu

https://www.routledge.com/Multispecies-Anthropology/book-series/MANE

Liang Shaoji's Silkworm Art

Untangling Multispecies Craft

Feixuan Xu

Routledge
Taylor & Francis Group
LONDON AND NEW YORK

First published 2025
by Routledge
4 Park Square, Milton Park, Abingdon, Oxon OX14 4RN

and by Routledge
605 Third Avenue, New York, NY 10158

Routledge is an imprint of the Taylor & Francis Group, an informa business

British Library Cataloguing-in-Publication Data
A catalogue record for this book is available from the British Library

ISBN: 978-1-032-91459-6 (hbk)
ISBN: 978-1-032-94133-2 (pbk)
ISBN: 978-1-003-56909-1 (ebk)

DOI: 10.4324/9781003569091

Typeset in Times New Roman
by KnowledgeWorks Global Ltd.

To my parents.

Contents

Figures

Acknowledgements

My deepest appreciation and respect go to Liang Shaoji. Without the artist's persistent energy to create and share, this book would not have taken shape. I am also greatly indebted to all my correspondents, who selflessly entrusted me with their knowledge, feelings, and stories during my fieldwork – some of whose names must remain unspoken here for ethical concerns. All the silkworms I encountered have left their invisible yet tangible traces, subtly woven into this project.

I am grateful to my PhD supervisor, Dr Zheng Bo; my doctoral examiners – Dr Alex Blanchette, Dr Damien Charrieras, and Dr Linda Lai; and the anonymous peer reviewers invited by Routledge. As the earliest readers of this book's formative versions, they offered not only perceptive feedback but also affirming trust in its value. I am equally indebted to Dr Paolo Fortis, Prof Robert Layton, Dr Elaine Gan, Dr Eugenia Kim, Dr Sarah Huang, and the 'Wanwu Practice Group', whose less formal yet incisive insights nourished my academic enquiries. The research for this book was supported by the Postgraduate Studentship from City University of Hong Kong, the Fulbright Program, and Asymmetry Postdoctoral Fellowship.

My family's unwavering love has been the cornerstone of my inner strength. The steady, patient, and caring company of dear friends has also sustained my calm through the years. My profound thanks extend to my intellectual shadow-advisor, Dr Magnus Wilson, the sharpest critic of my crude drafts, who has tirelessly – albeit intermittently – guided me for nearly a decade to write with clarity and dignity. His mentorship marks the completion of this book as a milestone in my scholarly journey. '*All's well that ends well*', as he aptly quoted.

1 Journey into Silkworm Artistry

'Do farmers still grow sweet clementines as a supplement to rearing silkworms?'

The question slipped out inadvertently, hanging in the air between me and Yian as we drove along the narrow country road. Yian, my longtime guide and friend in Village Shangping, picked me up from the coach station in town on this balmy day in late October 2024 – nearly five years since my last visit in 2019, just before the world had been upended by the COVID era.

'Nope, not anymore', Yian responded, his eyes sweeping across the vast farmlands. 'The sales of clementines were pretty poor last year. Villagers had to discard them on the ground as fertilizer. It's better to raise silkworms, isn't it? You can see the money in just a dozen days. But it takes three years to grow fruit trees, and who knows what will happen in that time'.

The mulberry trees stood sturdy beside us, their branches still dotted with scattered leaves, almost translucent in the generous sunlight. It was a scenery I had grown familiar with during my research into silkworm husbandry and silkworm art.

'What about the purchase price of cocoons now?' I asked Yian, recalling that in 2019, the average price in this county Yizhou had fluctuated between 18 and 27 yuan per catty of unprocessed cocoons throughout that year, disappointingly low due to an unprecedented freezing spring and an excessively humid June.

'Last year, the price wasn't bad, averaging 26 yuan per catty. This year, much better, around 28 to 29, and the latest batch this month went for 32 to 33 yuan. Farmers raise a batch of silkworms in two weeks, and then earned 8,000 to 9,000 yuan. It's a good deal, really. Worth the gruelling work'.

Reliable cash worms – this is how sericultural practitioners might depict domesticated silkworms. Yet, my fascination with this creature arose from a far more otherworldly and less utilitarian scene. In 2017, as a first-year doctoral student curious about the landscape of ecological art in China, I was leafing through a stack of heavy hardcover art catalogues borrowed from university library. It was during this contemplative browsing of the catalogues that I first encountered the mesmerising images of Chinese artist Liang

DOI: 10.4324/9781003569091-1

Shaoji's installations – a moment that would fundamentally remould my understanding of silkworms. Here in Liang's oeuvre, these creatures appeared to mingle with the artist behind the art-making scenes, their identity as mere agricultural resources fading into something far more nuanced.

Liang Shaoji, born in 1945, is a contemporary artist who has spent nearly 30 years raising and working with domestic silkworms (*Bombyx mori*) from 1989 onwards. The '*Ziran* Series', often translated as 'Nature Series' in English, is representative of his silkworm art, renowned worldwide for its embedded Chinese, or at least East-Asian cosmology and conception of nature. Liang's silk-bound installations, video works, and photography have been exhibited in domestic and international art venues such as the Hayward Gallery (London, 2012), ShanghART Gallery (Shanghai, 2014), Al Riwaq Art Space (Doha, 2016), MWoods Museum (Beijing, 2018) and the Smart Museum of Art (Chicago, 2019). He has also participated in Biennales in Venice (1999), Lyon (2000), Shanghai (2000), and Pančevo (2002). In 2009, Liang was awarded the Prince Claus Award under the annual theme 'culture and nature', an international award to honour artists and cultural practitioners contributing to cultural development. Liang's art practice is celebrated as offering 'a meditative approach in which art becomes nature and nature becomes art' in the award committee report (Prince Claus Fund 2009).

Liang named his main series of artworks '*ziran*', a concept he has frequently talked about, pondered over, and attempted to present via art and has been acclaimed for in the art world. Not surprisingly, Liang Shaoji's silkworm works are often categorised as 'ecological art', as his projects involve non-human partners, the silkworms, and on the one hand, are interpreted by art critics as dealing with human relations with nature – a point I will return to discuss in detail later. Liang's artworks have been juxtaposed with other works conventionally categorised into eco-art, bio-art, or artefacts in group exhibitions under titles such as 'Longing for Nature, Reading Landscapes in Chinese Art' (Museum Rietberg, 2020), 'Growing' (Chronus Art Centre, 2019), and 'Trees of Life: Knowledge in Material' (NTU Centre for Contemporary Art Singapore, 2018). The adaptability of Liang's works to be positioned in versatile contemporary art contexts is partly because the triangular relationship between the artist, the material (silk), and the organism (silkworms) complicates the appreciation and reception of Liang's silkworm art.

This hermeneutic ambiguity was evident when I attended the opening of Liang Shaoji's solo exhibition 'As If' at MWood Museum in Beijing (14th September 2018) where I encountered two photographs of his installation 'Snow Cover' (雪藏) on site. 'Snow Cover' contains many discarded objects, such as an empty coffee carton, a telephone, a cell phone, an integrated circuit board, some poster paper, and a plastic cup, all scattered on a wooden board. The everyday objects on the flat board are covered under a full sheet of white and translucent silk threads, with entangled filaments discernible and shimmering under the light (Figure 1.1). I arranged an interview with the artist

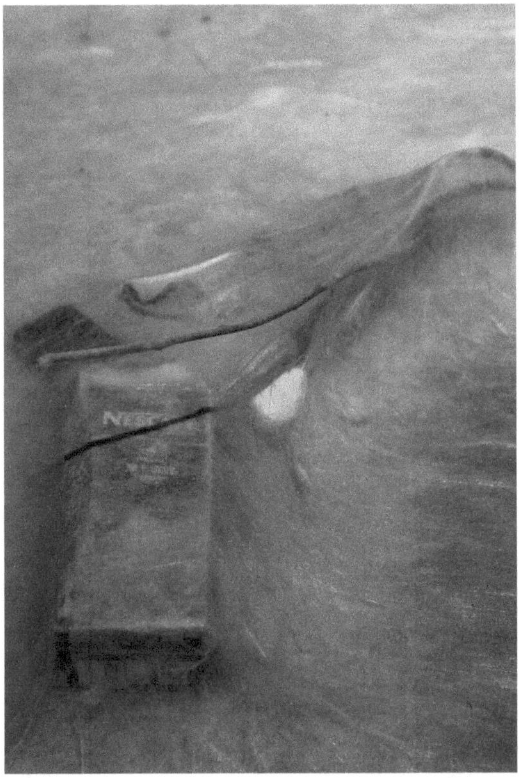

Figure 1.1 'Snow Cover No. 2', 2013–2014. Copyright: Liang Shaoji and ShanghART Gallery

two days after the opening when Liang Shaoji and I talked on a couch in the museum's office space.

I asked about 'Snow Cover', one of my favourite among his silk-bound installations: 'Some critics I read got a sense of "covering-up" and "erasing" in this work (von Drathen 2011; Xia 2015), but what I discerned was more towards inclusion, tolerance or even mercy. What do you think?' As those art critiques unpack the ethereal visual composition of 'Snow Cover' and its possible conceptual connotations, my anthropological training seemed to kick in. I perceived a vaguely shamanic spark in this work and therefore wanted to know the artist's explanation.

Liang's intonation was always passionate when talking about his works: 'I would rather say it's about "healing". The "flat cocoons" spun by the silkworms are light, white and ethereal, like a handful of snowflakes. Silk as a material contains proteins, one of the most fundamental biomolecules comprising life forms and it serves as a treatment in this installation to re-provide you with the key element of life. The winter snows, on the other hand, are nourishment for the land, and under the snow cover is not death but endless transformation and hope'.

It was a poetic exposition where an analogical association was made between white silk and white snow through their material and functional characteristics. The concept of 'healing' crept back into my mind from time to time when I later explored more into the 'flat cocoon' technique employed by the artist to create most of his silk-bound installations like 'Snow Cover'. Healing has 'religio-magical' undertones especially in anthropological contexts when its enactments involve body, mind, and spirit and emphasise process, embodiment, sensation, and aesthetics, thus regarded as the litmus test for the ritual efficacy. Liang's statement about the implication of 'Snow Cover' has triggered more inquiries in this book. What was it like for the silkworms to crawl over the human-used material such as smart phones and secrete 'flat cocoons' on them rather than the habitual ellipsoidal cocoons? What happened to the silkworms and the artist in the processes to complete those large-scale installations? Does 'healing' in 'Snow Cover' make sense only on a symbolic metaphorical level, as a wishful projection on the silk and silkworms, or is healing also enacted on a ritualistic if not a functional level? 'Snow Cover' became the entry point for my research project on Liang Shaoji's artistic collaboration with domestic silkworms.

Meanwhile, when attempting to understand '*ziran*' and its associated culturally particular natural cosmology in Liang Shaoji's art practice, I realised that art critics focusing on iconographic, aesthetic and ethical qualities of those artworks fail to address the generative relations between silkworms and the artist in the art-making processes and the vital role of silk as a physical, functional, and metaphorical medium (e.g. Gladston 2011; von Drathen 2011; Xia 2017; Yang 2018; Zulueta 2016). Moreover, the physical and symbolic continuity of silk-silkworm combo in Liang's projects destabilises the boundaries between body, organism, object, and material, thus challenging the analytical categories of bio-art, animal art, eco-art, and conceptual art in contemporary art criticism. If provisionally translating '*ziran*' with Daoist connotations as merely 'natural', the questions still follow as such: Is silk a natural material to be employed? Does the generative alliance between silk and silkworm count as a natural relation? Does the artist work as part of nature or work on nature? Does '*ziran*' (natural or harmonious) appropriately describe the ways Liang experimented with silkworms to create art as claimed by both the artist and art critics in most of his projects? After all, how do human and non-human organisms, natural and non-natural materials, and artistic techniques in this trans-species interaction determine the efficacy of Liang Shaoji's silkworm art and the expressive power of the spirit '*ziran*', physically and metaphorically?

To address all these questions, this book adopts an anthropological perspective to comprehend Liang Shaoji's art and the intricacy of '*ziran*', a culture-bound but vaguely defined concept related to ecology and nature, being re-constructed in Liang's creative practice. I investigate Liang's artistic techniques of cultivating silkworms and making silk art, with a focus on the functional-spiritual role of the silkworm as the organism and the silk as the ritualistic-material object. Based on multi-sited fieldwork in China with the artist, silkworm farmers, and sericultural scientists, I re-examine the material composition and symbolic articulation of '*ziran*' (natural, spontaneous, harmonious, 'so of itself', 'just how it is', non-assertive, etc.) in Liang's silkworm art, which is often over-simplified as a Chinese Daoist-influenced egalitarian cosmology of nature or less-manipulative approach to animals by critics, as I will briefly review in the following sections. As the book upfolds, my ethnographic experience will illuminate what counts as *'ziran'* and its anthropogenic manipulation complicated by Liang's artistic experimentation.

Eco-art and Bio-art in Art History

There are two main critical approaches towards Liang Shaoji's art in the art history framework, placing his silkworm projects in either the theme-led category 'ecological art (sometimes eco-art)' or the method-led 'biological art (also known as bio-art)', in face of current environmental destruction and the anthropocentric paradigm (Demos 2016; Weintraub 2012). As for the former, Weintraub's book (2012) offers a panoramic overview of ecological art. As a category that can be defined by artworks' strategies, issues, and materials, ecological artists often deal with topics such as sustainability, climate change, biodiversity, mass extinction, and catastrophism via methods such as paintings, video, sculpture, performance, social-engaged practice, digital art, and design. Recent decades have witnessed more diverse artistic practice under ecological themes and art criticism towards these works has experienced an interdisciplinary trend where theories in social sciences and other humanities have been brought into the discourse of art history. Demos (2016), for example, focuses on the aesthetic and political engagement of ecological art in his monograph 'Decolonizing Nature: Contemporary Art and the Politics of Ecology'. The politics of ecology is concerned with the regional power asymmetry in globally distributing the costs and benefits of environmental exploitation, and consequential poverty, racism, and neo-colonial violence. The political-ecological approach acknowledges the potential power of artistic intervention for environmental justice and thus, as Demos (2016) suggests, activist-artists should take more ecological commitments in their practice. These analytical frameworks, however, are not fully applicable to projects that are less politically participative or socially engaged, instead addressing more conceptual or aesthetic aims under the umbrella term of eco-art.

When it comes to biological art (or bio-art), this can be ecological when it tackles multispecies interaction, nature-culture dualism, or symbiotic biosphere. As the human-nature relationship has been one of the major motifs in contemporary art, nonhumans are familiar 'guests'. Animals and plants are commonly employed as conveyors of symbols in human narratives, such as metaphor, myth, and folktale. They also serve as visual prototypes of images, either via direct representation or virtual parody. Bio-artists, however, seek a less exploitative and more egalitarian way to collaborate with other life forms. They make use of the processes of life, the developments of a micro-organism, or the evolution of species to realise their projects (Baker 2000; Eisenman 2013; Jevbratt 2009; Kirksey and Helmreich 2010). The recent development of media and biotechnology parallels a material shift in the manipulation of non-human matters, where dead body parts, DNA, tissue and living species, etc. are utilised by bio-artists to sculpt or mould uncanny living entities, often with the assistance of biologists (Stracey 2009). Artworks such as Eduardo Kac's GFP (green fluorescent protein) glowing bunny 'Alba' (2000) and Damien Hirst's exemplary 'Natural History' series sculptures with specimens preserved in formaldehyde in glass tanks reflect upon the uncomfortable and uncanny relational identities across species today when humans both 'kill and consume, protect and love' animals (Eisenman 2013: 260).

Meanwhile, as a scholarly response to the fast-developing biotechnologies against the background of structural inequalities and devastation of ecological systems, some critical theorists (often conveniently labelled as 'posthuman' for their anti-anthropocentric stance) have challenged the central position of the humanist ideal of 'mankind'. Instead, they have promoted new terms to capture the ambiguous liminality between flesh and machine, wetware and software, human and animals, etc., thus hoping to eliminate the species hierarchy and eco-injustice penetrated both the discourse and practice (e.g. Barad 2003; Braidotti and Hlavajova 2018). In regard to the nonhumans, the posthuman scope not only involves those unattended marginalised organisms beyond the human-centric realm but also the 'technologically manufactured others' (Braidotti and Hlavajova 2018: 2). In Haraway's ground-breaking essay 'A Cyborg Manifesto' (1985, reprinted in 1991), the feminist sensitivity is mingled with sci-fi tones in the narrative of the 'cyborg' (blended from 'cybernetic organism') where the ideological boundaries become mutable and strict dualisms are dismantled. Similarly, Bennett's (2010) extension of the Deleuzian idea 'bodies without organs' to the geological level argues that the earth is filled with vibrant matters and all entities on this planet are embodied with omnipresent but unnoticed vitality, thus demolishing human exceptionalism. The thriving posthuman literature (e.g. Bennett 2010; Haraway 2008; Latour 2017) has been largely adopted as conceptual tools in the analysis of bio-art (Bakers 2013; Da Costa and Philip 2008; Kirksey 2014). However, the innovative vocabularies from critical theories, on the one hand, more accurately describe the messy hybrids or generative relations that have emerged in many ecological or biological art projects, but they add little insight into the hermeneutic understanding of artworks themselves.

To evaluate artworks involving living organisms, whether silkworms or other animals, the existing eco-art framework in art history focuses on ecological themes in association with social, cultural, economic and political factors in each case, while bio-art criticism, besides looking at the final presentation, also examines the art-making processes between artists and nonhumans. Discussions on the backstage operations in bio-art projects tend to focus on the procedures regarding the ethics of art creation, especially animal rights or welfare (e.g. Baker 2000; Jevbratt 2009, Yang 2018), rather than how the ethical evaluation contributes to the meaning construction of the whole artwork. Moreover, as bio-artists play with the conceptual instabilities of life, body, autonomy, and nature by tactically borrowing biotechnologies which are embedded in a homogeneous scientific discourse, there is a tendency for aesthetic and cultural particularities to become secondary, if not unremarked, in the appreciation of bio-art; this discursive hegemony might risk missing the nuances in non-scientific local contexts in artworks, although it is not the case in art critiques on Liang Shaoji (details see below). Before further examining how Liang Shaoji's silkworm art has been appreciated and interpreted as either bio-art or eco-art within art criticism's disciplinary framework, I will first provide an overview of ecological art in contemporary China where Liang Shaoji's art can be culturally grounded, and then briefly explore the 'Chineseness', especially Chinese-specific 'ecological ideas' manifested in those artistic practices.

Ecological Art in Contemporary China

Chinese contemporary artists, like their non-Chinese peers around the globe, have expressed their concerns for environmental issues discursively, visually, and affectively and have also offered innovative propositions and imaginaries of an interspecies symbiotic future. A special issue of the *Journal of Contemporary Chinese Art* edited by Zheng and Lee (2016) collected essays on East Asian ecological art practice, including painting, performance, documentaries, installations, and socially engaged art. The editors summarise four reservoirs of ideas in the Anglophone academic literature on art and ecology, namely, scientific studies, Western philosophy, indigenous cosmologies, and climate change activism (e.g. in Demos 2016); hence they intend to add a fifth resource, East Asian thought, as a crucial anchoring point to explore eco-art in East Asia. Daoism ('The Way') featuring the relationality, cyclical pattern, and immanent vitality in East Asian metaphysics (Perkins 2015 quoted in Zheng and Lee 2016: 217) has been emphasised as the underpinning of East Asian eco-art. Essays in this issue provide case studies of Chinese ecological art practice such as Xu Bing's educational 'Forest Project', Liang Shaoji's silkworm art (Zulueta 2016), and Gu Dexin's installations containing raw meat.

Nevertheless, it is worth noticing that despite the limited volume of literature delineating a holistic picture of eco-art in China, existing studies in art history alongside ecocriticism tend to emphasise Chinese particularities, namely, cultural, religious, cosmological, or ethical features in art projects from this region. Chu (2015), for example, investigates the aesthetic pastiche and parody of Chinese traditional landscape painting '*shanshui*' in contemporary visual art including animated cinema and experimental performance. Although Zheng and Lee implicitly admit that art from East Asia including China does not necessarily refer to East Asian metaphysics, they continue to argue for 'localness', that is, the ecological thought exclusive to the region as an entry point of analysis as part of post-colonial commitments to build multiple discourses in eco-art criticism (2016: 219–220).

In broader discussions on the 'Chinese-ness' of Chinese contemporary art, it is contested whether art projects by Chinese artists are necessarily rooted in Chinese cultural tradition or as Tsao and Ames (2011) suggest, this may be 'yet another excellent example of cultural self-colonization' (xv). Artworks touching Chinese experiences and issues or reflecting the artists' early training and fields of activities sometimes need to be identified as 'Chinese', and with this label of 'otherness', they can gain recognition in the international art arena as an alternative of Western contemporary art (Tsao and Ames 2011: xiii). Liu (2011) expresses similar worries that Chinese artists through the categorical designation of national identity are expected to represent a unified, often invented Chinese cultural tradition or past (Hobsbawn and Ranger 1983 in Liu 2011: 143) and this stereotype 'downplays their own diverse backgrounds' (144).

Regarding the global recognition of Chinese contemporary art, some scholars see more positive sides of the interculturality in Chinese artists, especially those diasporic examples such as Gu Wenda and Xu Bing who refuse to play a facile role of the cultural 'other' (Isto 2015; Liu 2011; Vinograd 2011). Vinograd (2011: 96) proposes critics should not overlook the culturally based references but the international or transnational contexts those Chinese artists appear in are also worth exploring. These intercultural encounters, according to Isto (2015: 197), can shed light on how globalisation exceeds issues of cultural communicability, especially when differences between human and nonhuman go beyond the national or cultural level in face of planetary ecological challenges. Isto hence applies posthuman theoretical tools by Cary Wolf and Timothy Morton to analyse Xu Bing and Gu Wenda's bio artworks (2015) rather than resorting to Eastern intellectual resources as Zheng and Lee (2016) insist. Nevertheless, Zheng and Lee in their introductory essay present one of the keywords '*ziran*' in East Asian metaphysics, which is also the anchoring concept in Liang Shaoji's silkworm art. Before examining how '*ziran*' as a culturally specific ecological concept in Liang's artworks has been appreciated by critics, I will provide an overview of Chinese ecological cosmology as a background for further discussions.

'*Ziran*' in Chinese Contexts

In Zheng and Lee's (2016: 216) summary of East Asian metaphysics shown in artworks, '*ziran*' has been explicitly distinguished from 'nature' although the latter is mostly used as the former's English equivalence. The conception of 'nature' is based on the dualist premise dividing nature and culture, object and subject, bodies and thoughts, wild and domesticated, etc., which has now been intensely criticised and deconstructed for those problematic dichotomies associated with it (e.g. Latour 2004; Morton 2007). Strathern (1980), for instance, acutely claims that the dichotomies in symbolic categories between gift and commodity, men and women, domestic and wild, and natural and cultural are not cross-culturally pervasive around the world with the ethnographic support in her analysis of the gift exchanging practice in Melanesian communities. '*Ziran*', on the other hand, can be literally translated as 'just how it is' (Zheng and Lee 2016: 217) or 'so of itself' (Jenkins 2002: 42). Different from nature as an environmental resource to be mastered, '*ziran*' as a noun refers to an organic encompassing unity with the interdependent myriad things all-included, and as an adjective, it denotes the spontaneous, intrinsic, harmonious ways or patterns things naturally follow (Ames and Hall 2003; Jenkins 2002).

Chinese '*ziran*' is one of the key notions in Daoist and Buddhist *Chan* (*or Zen*) worldviews, which are often easily designated as Chinese, East Asian, or even Eastern ecological thinking and the artist Liang Shaoji (2002, 2008) frequently resorts to in articulating his artistic inspirations. Although natural cosmology in Chinese traditions, if not in Chinese religions, is another rich field of scholarship, here a brief overview will help to contextualise '*ziran*' in the larger picture of Eastern worldviews characterised by enduring harmony, a holistic unity of humanity and cosmos, as well as unbroken continuity, productive correlations and balance among all internal life forces (Cheng 1976; Tsao and Ames 2011; Vachon 1983 in Jenkins 2002). Chinese traditional ecological wisdom encapsulates intellectual resources from Confucianism, Daoism, and Buddhism, three bodies of doctrine which count as functionally equivalent to religions (Ching 1993) or as systems of moral teachings on social behaviours (Bodde 1991, quoted in Jenkins 2002). Jenkins (2002) divides Chinese ecological worldviews into two kinds, one delineated in elite thought from these three doctrinal lineages and the other manifested in popular utilitarian practice.

'*Ziran*' is prevalent in Confucian 'anthropocosmic social ecology' (Tu 1989), where it serves as a moral force of universal orders on which human rationality and moral perfection are based; in an ideal state, humans should take their agrarian and social responsibilities to seek a harmonic balance between human order and universal order (Jenkins 2002). Buddhist Mahayana traditions in China emphasise cosmic inter-relatedness and teach ethical sensibility and responsibility to all sentient beings. The *Chan* sect, under a great influence of Daoism, proposes respect for '*ziran*' and an intuitional and less ascetic approach to enlightenment (Wilber 1981 quoted in Jenkins 2002). '*Ziran*' is most prominent in Daoist philosophy, which had long been pervasive in Chinese popular thought in ancient times as Needham (1956) argues. Ames (2001), the English translator of the Daoist classic '*Daodejing*' and a comparative philosopher, summarises how Daoist philosophy prioritises process and change over form and stasis, situation over agency, historia and mythos over logos, narrative over analysis, contingent and negotiated harmony over deterministic and necessary teleology, a dynamic radial centre over boundaries. Daoist processual ontology acknowledges continuity and diversity, but on the other hand, does not sustain a progressive view of history nor conservation of the status quo (Ames 2001). Thus, there are two keywords associated with the conception of '*ziran*' in Daoism, '*qi*' the immanent vitality in '*ziran*', and '*wuwei*' the non-action principle towards '*ziran*', also interpreted as actions not violating universal orders to downplaying invasive human interventions (Needham 1956). This stance suggests that Daoist ethics respect natural patterns in nature with limited human interference in practice (Ching 1993). But before examining how contemporary artworks borrow, appropriate, or refer to this intellectual heritage, it should be noted that religious, textual and institutional Confucian, Buddhist and Daoist teachings on ecology should not be equated to popular, grassroots, and secular beliefs in ancient and contemporary China, which are more characterised by concerns for utilitarian efficacy as well as 'this-worldly' prosperity and immediate happiness (Jenkins 2002).

An unresolved contradiction of Chinese ecological thought lies in the disconnect between an idealised worldview and the actual situation. As Jenkins (2002: 47) points out, even though Chinese environmental ethics towards '*ziran*' worships reverence, compassion, and stewardship, it failed to effectively halt environmental degradation and an ecological crisis in modern China and thus becomes an ongoing issue how to effectively exert practical influence in academia, in public policy-making and the private sphere. Nonetheless, contemporary artists have certainly turned to natural cosmologies of ancient China when touching upon ecological issues, making conceptual innovation and reflection (if not necessarily practical action) especially in artworks that take non-activist approaches (e.g. Chu 2015). However, critics should bear in mind that the danger of resorting to 'tradition' even at a discursive and speculative level is that traditions can be invented or reinvented. In Liu's (2011) critique of Xu Bing's installation 'Where Does the Dust Itself Collect?' (2004), for example, the title is excerpted from a Buddhist *Chan* proverb, but Liu cautiously scrutinises whether the *Chan* idea Xu Bing adopts is decidedly Chinese variation on an antique Indian philosophy or a thin, popular and Westernised version of the sect. Although authentic allusion to an original or ancient ecological doctrines is definitely not a golden criterion for good artwork, a probe into which version of ancient thought artists have referred to, as well as how they appropriate it and why they choose to do so allows for a more accurate analysis of their artworks, and more importantly, helps to judge whether a work plays a hollow, superficial and airy transposition of concepts or instead brings genuine fresh insights to the art world or broader contexts.

Liang Shaoji in Art Criticism

Although Liang Shaoji is not an expatriate Chinese artist like Cai Guoqiang, Huang Yongpin, Ai Weiwei, and Xu Bing, the 'interculturality' of these artists (Isto 2015; Liu 2011; Vinograd 2011) is also manifested in Liang's personal experience and artistic creation. As the artist (2017) recalls, he studied at the Affiliated High School to China Academy of Art until 1965, an institution founded by pioneering Modern Chinese painter Lin Fengmian where Liang was exposed to and attracted by German expressionist art even though Soviet realism was dominant in art education due to the ideology in China at that time. The start of the Cultural Revolution in 1966 halted university enrolments nationwide and dashed Liang's ambition to study sculpture. Liang then was assigned to work as a fabric designer in a craft factory in Taizhou city in Zhejiang Province, which was one of the largest producers of woven products for the export market at that time. This practical experience equipped Liang with solid knowledge and sensitivity to various materials which later also fostered his silkworm art creations in middle age.

In the 1970s and the early 1980s, Liang was appointed as director of the government-founded Institute of Arts and Crafts of Taizhou and also worked at the Overseas Exhibition Office of the Light Industry Department of the Beijing Government for more than two years. These occupations allowed him to experiment with textile art and fabrics to gain experience in curating in large exhibition spaces and most importantly, to travel abroad on official business and visit galleries in Germany, France, the USA, and the former USSR, when postmodern art in the West was still in its heyday. In Zulueta's (2016: 382) view, these trips to foreign countries and the regained access to books of Western writers (Liang himself has frequently cited and quoted Nietzsche, Heidegger, and Agamben) enlightened the artist to critically think about the constraints upon Chinese society and himself as an individual after the Cultural Revolution.

It was a serendipitous opportunity that drove Liang Shaoji's career from arts-and-craft design in fabric art to contemporary art in his 40s. In 1986, Liang enrolled at the Varbanov Institute of Tapestry at the China Academy of Art in Hangzhou, Zhejiang mentored by the Bulgarian painter and tapestriest Marin Varbanov (1932–1989) known for his soft sculptures consisting of wool, silk, and bamboo textiles. Liang Shaoji's hands-on training and experience in soft sculpture and craft there during the late 1980s deepened his mastery of various materials such as hemp, stone, seaweed, and glass (Liang 1994). When Liang's artistic interest later narrowed down to silk and silkworms, his probe into '*ziran*' still revolved around the spirituality of materials. The installation 'Yi Series No.1: Magic Cube' (1988) heralded more than 30 years of silkworm art creation (Figure 1.2) composed of artificial silk, cocoons, *xuan* paper, linen, bamboo, and metal frames. In February 1989, 'Magic Cube' was exhibited at 'China/Avant-Garde Art Exhibition' in the National Art Museum of China in Beijing; a few months later the Tiananmen Square protests and massacre took place.

Chinese contemporary art emerged in the global art world as a category in the early 1980s when the Start Group anticipated the art movement 'New Wave of 1985' as a direct response to then social protests and political

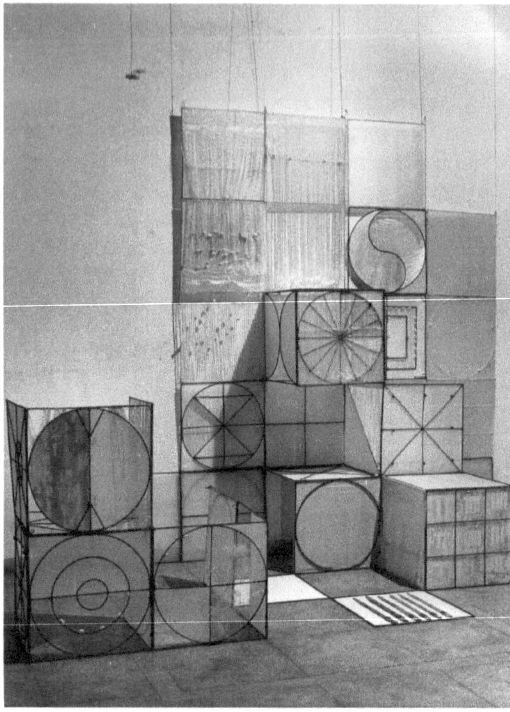

Figure 1.2 'Magic Cube/*Yi* Series No. 1', 1988. Copyright: Liang Shaoji

censorship of art works (Tsao 2011: 2). It was after the Opening of China policy initiated in 1978 that artists were able to draw on Western literary, semiotic and cultural theories in their art making (Vinograd 2011). However, the post-Tiananmen era saw a shift in Chinese contemporary art world towards post-modern parodies of Maoist propaganda in domestic art scenes. Meanwhile, cutting-edge artists in exile tended to strategically juxtapose multi-cultural elements to mainly non-Chinese audiences to gain tremendous fame in contemporary art market, which Bhabha criticises as showing a spurious posture of 'internationalism' as those expatriate Chinese artists live in 'the nations of others' (2002: 30).

After the 'China/Avant-Garde Art Exhibition' in 1989, Liang Shaoji stayed in China to continue his career as a contemporary artist. His 'Nature Series' started with an installation simply named 'No. 1' (1989) with loops of barbed wires in the shape of a globe entangled with several ellipsoidal cocoons when the artist was yet inexperienced with the 'flat cocoon' technique employed in his later works (Figure 1.3). Liang spent nearly ten years since then raising silkworms with village farmers in a suburb of

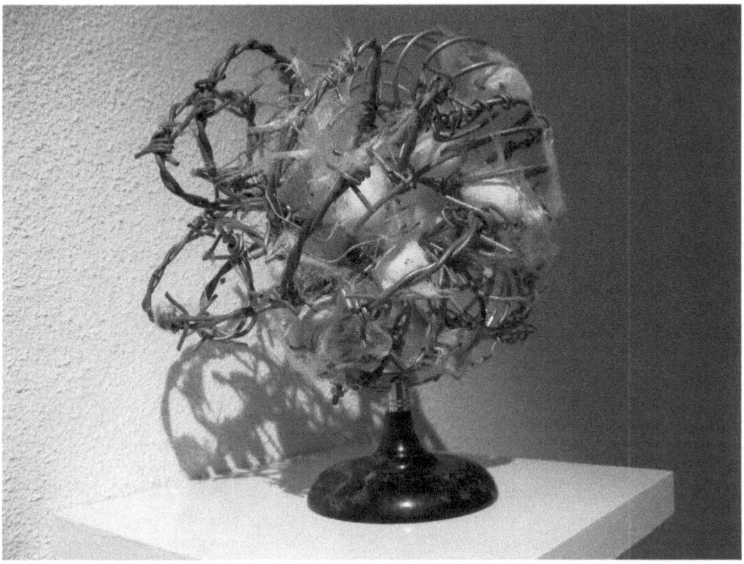

Figure 1.3 'Nature Series No. 1', 1989. Copyright: Liang Shaoji and ShanghART Gallery

Linhai County in Zhejiang and learning from sericultural experts in the then Zhejiang Agriculture University (merged into Zhejiang University in 1998). As the artist himself articulates (in Yang 2018), his initial artistic vision was to have silkworms secreting silk on metal planes as a metaphor for the ongoing transformation in China from a pre-modern agrarian society to a modern industrial information one. As he became familiar with the physiological characteristics of the silkworm and sericultural methods, Liang gained more insight upon the collaborative potentials between art and science and determined to make silkworm-art as his life-long art experiment.

Liang Shaoji is not the only Chinese contemporary artist to create art with silkworms. Established artist Xu Bing has made 'American Silkworm Series' since 1994 containing four installations where silkworms spun threads on objects such as a dictionary, Kafka's book 'The Metamorphosis' or a VCR player, as a reflection on materially expressed natural language and human language filled with cultural identities (Vinograd 2011: 110). Neri Oxman's 'Silk Pavilion I' (2013) creates a biomicritic silk-steel dome, highlighting the architectural and material capacities of silk cocoons, in collaboration with MIT Media Lab. More recently, Vivian Xu a media artist combines the capacity of the silkworm and cybernetic devices to build up a silk machine with biological and the computational logics entangled (2019). However, only

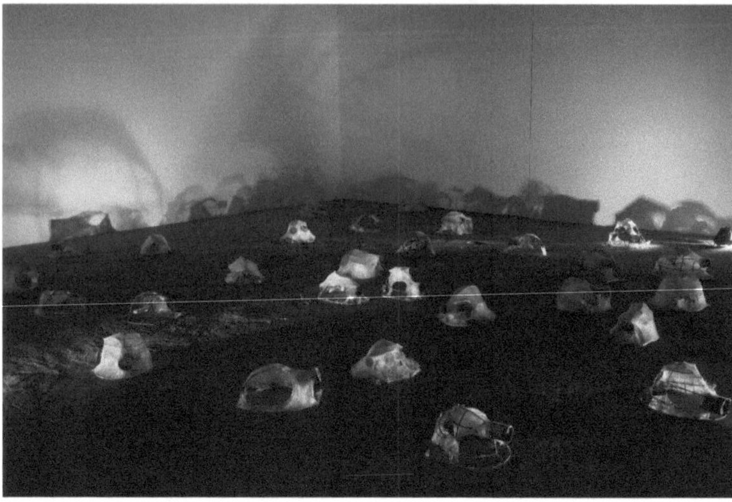

Figure 1.4 'Helmets/Nature Series No. 102', 2004. Copyright: Liang Shaoji and ShanghART Gallery

Liang has spent nearly 30 years maintaining a long and deep relationship with silkworms since the 1990s and keeps exploring the '*ziran*' motif throughout his art career.

Gaining more biological knowledge on silkworms over the years, Liang finished his early work 'Beds/Nature Series No. 10' in 1993, an installation which was later exhibited at the 48th Venice Biennale in 1999 where silkworms managed to secrete silk threads in the form of the flat cocoon and wrap charred engine coils in bed-frame shape. The flat-cocoon technique requires the silkworms to secrete silk filaments on two-dimensional flat planes as thin flakes rather than spinning commonly seen oval-shaped cocoons. The dominant compositional grammar of flat silk flakes covering objects in his silkworm art appears wholly in his later installations such as victim miners' helmets and headlights in 'Helmets/Nature Series No. 102' (2004) (Figure 1.4) and fossilised wood pieces dating back to the Tang Dynasty in 'Heavy Clouds' (2014–2018) (Figure 1.5) or partly in performances such as silk-wrapped naked female bodies in 'Self-roped/Nature Series No. 31' (2000) and mirrors in the performance 'Cloud Mirror/Nature Series No. 101' (2007). Other projects going beyond the prominent flat-cocoon technique in Liang Shaoji's silkworm art include the installation 'Broken Landscape' (2016) where the lifespan of numerous silkworms was spent and recorded on a long yellowish rectangular piece of silk fabric and the insects dispersed their faeces, urine, and eggs on the textile whose physical traces are kept in their original forms as little black dots and pale brown marks. Meanwhile, in the performance 'Listening to the Silkworm/Nature Series No.

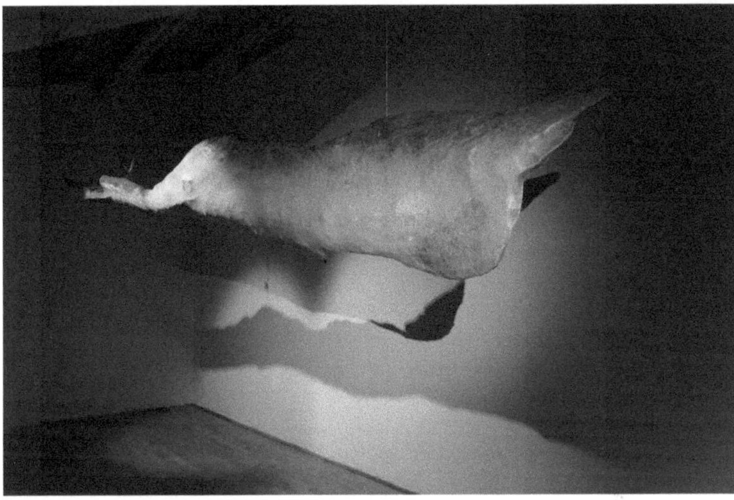

Figure 1.5 'Heavy Clouds', 2014–2018. Photo by author

96' (2006), a sound art installation invited the audience in the gallery to wear headphones and listen to the voice recordings of silkworms' chewing mulberry leaves and also spinning silk filaments.

Art critics have tended to conclude that Liang Shaoji's silkworm art expresses the artist's philosophical contemplation and social concerns, which outshines the aesthetic value and decorative manoeuvre in his practice. By juxtaposing soft raw silk and metal or plastic daily objects as the conveyors of metaphorical analogies, the artist is said to create a confrontational tension between environmental catastrophe and progress and development (Gladston 2011; Yang 2018). In particular, Liang's works deal with the uneasiness between the idealistic premodern nature and the excess of accelerated industrialisation (von Drathen 2011; Xia 2015; Zulueta 2016), as well as individual freedom and institutional violence during social upheavals, namely, in the 'Self-roped' performance (von Drathen 2011; Xia 2017). On a more metaphysical level, as the award committee report of the Prince Claus Awards (Prince Claus Fund 2019) suggests, Liang's artworks create evocative meditations on nature and human existence and offer an investigation into the ethics of the human relationship with nature.

Most scholars and critics tend to attribute the presentation of '*ziran*' and its associated worldview in Liang's oeuvre to a Chinese-specific, Daoist-influenced, and Eastern aesthetic of ecology, as frequently claimed by the artist in interviews and his own published notes (e.g. Brouwer 2007; Liang 1994, 2002; Xia 2015, 2017; von Drathen 2011; Yang 2017, 2018). The ecological aesthetics manifested in Liang's silkworm art conceptualises nature as an all-encompassing unity (Brouwer 2007), with the worship of ethereal and ambiguous art expression (Xia 2017: 61), and the embrace of the continuity between humanity and the divine nature (also known as '*tian-ren-he-yi*') (von Drathen 2011) in line with the broad exposition of Eastern Daoist natural cosmology discussed in the previous section. However, one obvious problem with these aesthetic analyses is that the interpretation of Daoism-flavoured '*ziran*' as 'nature' in English ignores the polysemy of '*ziran*' in its original cultural contexts, such as the aspect of spontaneity in this concept. On this premise, the aesthetic characteristics in Liang's works as declared by these critics appear logically loose, haphazard, and superficial, and moreover, fail to explore the tensions between ancient intellectual sources and contemporary appropriations in Liang's case.

As a notable exception, Yang's (2017, 2018) two essays on Liang Shaoji (2017, 2018) also discuss '*ziran*' as the essential motif in his silkworm works but acknowledge the non-assertive, spontaneous, and harmonious connotations of this term. Liang's approach to working with silkworms, as Yang (2017) argues, is causing no harm and rather meditative, as opposed to the anthropocentric manipulating orientation towards animals in Western bio-art. Although silkworm domestication and husbandry embody human control and transformation of nature, Yang (2018: 125–126) suggests that Liang's art practice, under the help of sericultural experts, has followed '*ziran*', namely, the behavioural pattern silkworms intrinsically follow since the artist has minimised human intrusion in the art-making processes as the Daoist mantra '*wuwei*' (non-action) encourages. Yang concludes that Liang's silkworm projects show an Eastern-style dialogue between human and '*ziran*' in contemporary contexts where his interactions with silkworms are in accordance with the natural laws of the universe. This follows another Daoist principle '*wu-wu-gui-jian*' (no hierarchy among the myriad of things) and suggests that Liang's works reflect his deep compassion and respect for non-human life (Yang 2018). It is therefore plausible in Yang's view to unpack '*ziran*' in its rich Daoist grounding when viewing Liang's artworks in terms of ethics rather than aesthetics.

Nevertheless, Yang's assertion of the non-intrusiveness in Liang's use of silkworms, which serves as the foundation of her subsequent reasoning on Chinese-specific Daoist ecological ethics, is not firmly supported by facts about how the artist has worked with silkworms in reality (see below). On the other hand, despite Yang's concern with the art-making process, especially the trans-species morality underlying bio-art literature (e.g. Baker 2000; Jevbratt 2009), her ethical considerations concentrate mostly on the artist and thus to some extents are disconnected from the symbolic, functional and affective analyses of his finalised artworks. In Liang's case, the separation of ethical considerations and meaning in a contemporary art criticism can lead to a lack of accurate understanding of his silkworm art, as I will delineate in later chapters.

Other than '*ziran*' as a culture-embedded concept in Liang Shaoji's art from eco-aesthetic and eco-ethical perspectives, some art critics have expanded their analytical scope to the Chineseness in Liang's projects beyond Chinese ecological thinking (e.g. Cheng 2007, 2015; Gladston 2011). According to Liang's own statements, he cautiously keeps a distance from the 'silk road' discourse infused with Chinese official ideology and rejects the assumption that he intentionally manipulates Chinese cultural symbols (quoted in Yang 2017), even if silk and silkworms dominate his artistic work and despite acknowledging the long-standing Chinese material culture of silk and ancient literati paintings have enriched the cultural, aesthetic, and linguistic connotations in his art. Indeed, homophonic Chinese characters, for example, have been frequently employed by Liang (2008; 2014) to elucidate his projects, such as playing with words under similar pronunciation '*can/chan*' – silkworm (蚕), Buddhist Zen (禅), the sound of flowing water (潺) and being broken (残) when explaining the installation 'Broken Landscape' and the video work 'Illusion of Flow'. Cheng therefore regards Liang's deliberate choice of the domestic silkworm as expressing 'his filial piety to his fatherland' and furthermore, 'an attitude consonant with China's post–Cultural Revolution nationalism and the more traditional Confucian morality' (2007: 69). With such 'sino-centric sentiments', as Cheng suggests, Liang aims to create an artistic language to establish a new 'Silk Road'.

Gladston (2011) also focuses on Liang's Chinese cultural identity but associates Liang's semi-retreating lifestyle in Tiantai County away from cities and his use of natural silk as an echo of the literati in ancient China, who chose a country lifestyle away from the political centre as an expression of disillusionment with the regime or dissatisfaction with their career as officials in state government. The unsettled relationship between order and disorder in Liang's art, as Gladston contends, is an implicit dissent to the drastic political and social turbulence in the late twentieth century undergoing in China under the rule of the Chinese Communist Party. In Liang's conversation with Gladston (2011), the artist, too, implied that Daoism as part of the Chinese intellectual tradition, cannot be completely separated from Confucianism, but has been mutually integrated and protected and therefore it was almost impossible to use Daoism solely to break with the existing social constraints in China. In the May Fourth Movement, Liang recalls how Chinese intellectuals borrowed Western resources to break with tradition which in a parallel manner, he incorporated Nietzsche's philosophy in his work. In this sense, the anti-totalitarian stance in his 'Nature Series' has not been fully appreciated in the artist's own view.

A less culturally bounded approach to Liang Shaoji's silkworm art is to concentrate on the participating roles of the silkworm and the silk fibre or the morality of trans-species collaboration but in the analytical framework of bio-art (e.g. Fok 2012; von Drathen 2011; Zulueta 2016). In Fok's (2012: 99) overview of contemporary Chinese art involving animals, various species function as effective tools to replace or collaborate with humans in order to present socio-cultural issues or life-death motifs in art. Performances by some Chinese artists such as Yu Ji's 'Pet's Kiss' and Huang Yongping's 'Theatre of the World', as Yang (2017) suggests, are ethically problematic in terms of these artists' instrumental and anthropocentric control and manipulation of live animals. In these terms, although it is contested whether Liang's ways of working with silkworms are free of cruelty as Yang suggests (Yang 2017), some critics notice the degree of active agency of either the spinning silkworms or of the inorganic silk threads in Liang's artworks (e.g. Cheng 2015; von Drathen 2011; Zulueta 2016). According to Cheng (2007, 2015), the artist's creative agency works with the silkworms' agency to live and evolve, thus collaborating in the art-making processes across species boundaries using cocoons filled with life and significance.

Meanwhile, Zulueta (2016: 385) is more concerned with the conceptual expression of Liang's oeuvre and interprets his artistic project as advocating empathy and respect for non-human agencies such as the silkworms' as the basis to reflect upon globalisation and reconstruct identities and cultures. By referring to the field of animal study which encourages treating animals as beings with their own agencies, interests, and experience (Shapiro 2002: 332, quoted in Zulueta 2016: 378), Zulueta proposes to distinguish animals' biology and perspectives from the representations humans have projected onto them. In Liang's case, Zulueta notes that the artist has underlined both the agency of silkworms and the 'human/silkworm coevolution' and the entailed 'blending of nature and culture' in a shared history (2016: 384). However, these broad examinations of the artistic contributions by agentive silkworms or silk filaments in Liang's art projects share a common deficiency that agency in the critics' reasoning has not been clearly defined and remains under-analysed. For instance, it is vague whether the agency is conferred to or inferred in the animal or the material, and moreover, the subject of agency also tends to change from silkworm to silk and sometimes even to the noun 'nature'. The arbitrary distribution of agency is evident in statements such as 'nature [...] act[s] as the artist' (von Drathen 2011) and 'the egg-shaped cocoon is [...] filled with vitality and meaning' (Cheng 2007: 69). As these scholars offer no further elucidation to clarify vitality or agency in certain theoretical frameworks, it is thus almost impossible to evaluate and relate their claims to other literature.

Overall, studies of Liang Shaoji's silkworm artworks have explored the aesthetic, conceptual, ethical, and cosmological aspects of both the art-making procedures and artworks' final presentations. Although the categorisation of artists or artworks is only a tentative exercise to facilitate discussion and parallel comparisons, critics with more concern for the ecological aesthetics and ethics delivered in Liang's works have tended to stress '*ziran*' or natural cosmology in a Chinese or East Asian context, despite the concept too often being reduced to 'nature' in English and detached from its original Daoist flavour. Trans-species morality in artists' working processes with other organisms is also the focus of bio-art critics' attention, where they tend to use Anglophone conceptual tools from posthuman theories such as agency rather than East Asian philosophy as an intellectual resource. Similarly, as bio-art as a category intrinsically suggests an interdisciplinary collaboration between art and biology, scholars who analyse Liang's approach to silkworms within the scope of bio-art refer only on the artist's consultation with silkworm biologists but ignore the collaborative processes between Liang and silkworm farmers, thus missing the larger context of Chinese sericultural production behind his oeuvre. Current art critiques of Liang Shaoji remain uninformed about how his interdisciplinary or trans-species collaborations occur, with the result that any elucidation of '*ziran*' or 'agency' in his artworks lacks detailed support. In the scope of contemporary art criticism, the symbolic connotation, aesthetic presentation, and ethical evaluation of Liang's artworks overshadow his artistic technique and the efficacy of his projects. This book hence turns to analytical frameworks in anthropology to better understand the interactive relations among the silk, silkworms, the artist, and the audience, whose processual entanglements are believed to dictate the meaning of '*ziran*' in Liang's art, as this book will mainly demonstrate.

Theoretical Framework: An Anthropological Approach

Art criticism and anthropology were once distinct. Contemporary art, in particular, can still be regarded as opposed to the now-obsolete category of 'primitive art' or 'aboriginal art', once central to anthropological studies of art (Firth 1992). As such, art traditions in non-Western small-scale societies were assimilated into a 'monolithic and evolutionary scheme' of Western art history (Layton 1991: 2) under the label 'primitive' in early ethnographic accounts and exhibited internationally in art institutions serving for joy or edification of privileged spectators (Ingold 2013). So, although terms such as 'primitive art' and 'aboriginal art' have been cautiously discarded and criticised in anthropological studies of art for their ethnocentric colonialist stance (e.g. Layton 1991), 'art' itself as a category remains unstable and contested in anthropology.

The concept 'art', as originated in western traditions, encompasses aesthetics and communication as two sets of values (Layton 1991). In art history, as Morphy (2013) summarises, objects can be counted as art either because they are institutionally acknowledged as repositories of values for both international academia and the art market, or because they occupy some aesthetic or interpretative properties, or even because the artists intend to make them as art. However, none of these criteria are fully suited to cross-cultural application in anthropological studies of art. Anthropologists explore art objects and practice in a given cultural context during both daily activities and special occasions such as rituals and ceremonies, where the category 'art' differs from most other classifications of artefacts in that there are no minimum functional (e.g. boat or headdress) or material (e.g. pottery) criteria that would exclude things from being 'art objects' (Ingold 2013). Apart from examining iconographic and aesthetic properties such as artistic form and style as art historians do (e.g. Layton 1991), anthropologists concentrate on sensory and functional aspects of art objects or processes, such as for presentational or representational purposes (Morphy 2013: 655). Anthropological studies of art have shown the entangled relationships between worldviews and everyday life with exemplary cases in non-Western societies such as body art (Gow 1999), decorated canoes for sailing (Munn 1977), and wood carvings denoting the growth of humans (Fortis 2010). The research scope of anthropology foregrounds art as an experiential category and thus destabilises the previously conventional distinction between art and artefacts by questioning how the ways of making objects, perceiving the environment, and knowing the world connect together through the lens of art. Anthropology also provides analytical tools to study art as mediator among humans, non-humans, and the landscapes in specific cultures and thereby complicate established ethnocentric notions such as ecology, materiality, skill, personhood, and sociality, as I will elaborate later.

Though art objects in small-scale communities have long been the primary focus in this discipline as an attempt to 'provincialize art history' (Kisin and Myers 2019: 238), there have also been anthropological studies on contemporary art and folk art with more focus on the social, cultural and economic matrices of art production and circulation among the theoretical tides of postmodernism, decolonisation, and globalisation emerging from the mid-1980s (e.g. Kisin and Myers 2019; Marcus and Myers 1995; Steiner 1994; Schneider 2017). Art has been entwined into processes of cultural identity constructions, copyright and cultural heritage legislation, and local or global political activism in contemporary societies (Morphy 2013), and on the other hand, has initiated more transdisciplinary collaborations in academia. The art exhibition with a published book 'Multispecies Salon', for example, is a pioneering curatorial project on cross-species encounters in the convergence of contemporary art practice, bioscience, and anthropology, where the boundary between making art and doing ethnography becomes blurred in both practice and theorisation (Kirksey 2014). In the expanding scope of art, conversations between art and anthropology take multiple forms and directions. Artists might conduct fieldwork as critical art practice while scholars draw on anthropological theories to unpack how art initiates and mediates social relations among people and things.

Anthropology, Art, and Agency

Art entered the mainstream of anthropology, as noted above, with two theoretical trends. Material culture studies were pursued in line with culturally oriented archaeology (e.g. Miller 1985; Morphy 2013) and, on the other hand, structuralist approaches influenced by semiotic linguistics (e.g. Leach 1973; Turner 1967) examined aesthetic, iconographic, and functional aspects of art (Morphy 2013: 655). The semantic or communicative features of art are often manifested in anthropological studies of exchange, myth, magic, ritual, and religion. The enchanting efficacy of admired artworks is not purely a result of their sensory (or aesthetic) characteristics. The effectiveness of art can be central in the interpretation of religious, shamanistic, or ritual events, but art often 'appropriates [its] function' today as the functional aspect is always auxiliary for the hermeneutic purposes (2013: 650). The functional properties can be relevant to the meaning of artworks but are never the main reason for including them in the art category. In the contemporary art world, more especially, one of art's functions is as a semi-commodity to mark values, which has little to do with the understanding of the artwork itself.

A more radical theoretical pathway to analyse art's efficacy in social relations, which is less tightly attached to the symbolic aspects of an artwork, has been proposed by British anthropologist Alfred Gell (1992, 1998) in his action-centred theory. Gell (1992) takes a strong anti-aesthetics stance in his essay '*The Technology of Enchantment and the Enchantment of Technology*' where he opposes the religious attitude towards sacred art and instead proposes atheism in the anthropological approach to art, with indifference to aesthetic values. Anthropologists should concentrate on 'the [art] object as an object', regarding it as a social being rather than a carrier of symbols or a beautiful thing, and instead scrutinise how this physical entity works to 'cast a spell over' a person both in art production and social processes (1992: 43). Unpacking the ethnographic details of Trobriand canoe handicraft, Gell contends that the complex and bright designs of canoe prows can captivate the viewers and motivate them to offer more valuable goods than intended in Kula exchanges. As he notes, art as a 'technical system' demonstrates 'a certain technically achieved level of excellence' (1992); this proficiency in skill guarantees the magic and technical efficacy of art in rituals and other ceremonial and commercial exchanges, whose social functions are enacted by mediating and creating social relations between the artist and the audience.

Gell's posthumously published book '*Art and Agency*' provides a more systematic theory to analyse art objects as distributed agents in social relations during the processes of 'art production, circulation and reception', rather than an isolated 'evaluation of particular works' (1998: 2). Following British social anthropology's traditional focus on social relationships, Gell (1998: 7) refutes an aesthetic approach towards art instead placing it in its social contexts; in this way, no prior definition of art is needed to satisfy aestheticians, philosophers or art historians, because the intrinsic nature of art lies in its relations with others. As Gell did not have time to revise this manuscript in his lifetime, some obvious flaws in his art theory have been severely criticised. As Gell's aesthetics-free study of art in Marquesan ritual or magic manifests little indication of 'artness', it can be seen as merely ritual or magic research (Morphy

2009). Bowden expresses similar concerns that there is no coherent definition of art in Gell's theory although he has indicated that art objects show 'technical expertise', 'imagination of a high order' and knowledge about 'the intrinsic mechanisms of visual cognition with subtle psychological insight' (Gell 1998: 68 quoted in 2004: 323). These artistic properties proposed by Gell, nonetheless, are not exclusively in artworks but also manifest themselves in entities such as advertisements and computer games. Furthermore, when Gell foregrounds the complexity of artistic techniques, he pays little attention to the 'conceptual originality' of art in revealing the world's complicatedness (Gell 1998: 322), which is an aspect of art aspect prominent in contemporary works.

Despite these drawbacks, the advantage of Gell's theory applied in viewing Liang's silkworm art lies in its facility to involve silk and the silkworm into interpersonal sociality in art-related situations without having to render animistic agency to them by causal inference (cf. Latour 1999). This book does not intend to completely abandon the search for aesthetic, sensory and symbolic aspects in Liang's projects contextualised in Chinese sericulture and silk culture. However, drawing on Gell's agency and art nexus framework can amplify the social relations in the vicinity of physical art objects. The complexity of artistic skills, as Gell (1992, 1998) stresses, guarantees the art's enchanting effect to motivate the viewers in rituals. In Liang's case, how Liang's art techniques affect the entanglement of the human artist, the audience, silkworms, silk, and other matters serendipitously touches upon the central notions in anthropology of art, such as agency, cosmology, materiality, skill, and efficacy.

Fieldwork Methodology

This book adopts qualitative anthropological methods to understand details of the art-making processes, the artist's artistic techniques, and his quotidian interactions with the silk and silkworms. Multi-sited fieldwork in reference to multispecies ethnography (ME) has been conducted as the main research method to facilitate further discussions on the expressive, presentative, and functional effects of *'ziran'* in Liang Shaoji's silkworm art. My fieldwork in mainland China lasted from early March to the end of November in the year 2019. The locales include Hechi and Nanning in Guangxi Province, Shaoxing and Taizhou in Zhejiang Province, Shanghai and Beijing. As this section shall outline, my investigation into how Liang Shaoji raised silkworms and made art with them mainly relies on participant observation as the cornerstone method in conventional fieldwork with complementary interviews. My interviewees involve not only the artist but silkworm biologists, silkworm egg breeders in factories, farmers and officers in sericultural departments, etc., for the purposes to gain extensive knowledge about silkworm husbandry in China as the larger practical context of Liang's art projects. This project also employs 'multispecies ethnography' (hereafter as 'ME') as both a new model of research and a new genre of anthropological writing (e.g. Kirksey and Helmreich 2010) to spotlight this insect from a less anthropocentric perspective. I lived and laboured with silkworm farmers as an apprentice to acquire practical sensory knowledge about silkworm husbandry and human-silkworm cohabitation. ME allows me to take notice of common activities and body forms within human-silkworm assemblages in certain landscapes (e.g. Tsing 2013: 32), which I will further explain below.

My pilot field trip to Beijing dates to mid-September in 2018. It was at the end of my first year when my initial doctoral research plan was to study ecological art practice in contemporary China. I attended the opening of Liang's solo exhibition '*Huang*' (As If) at MWoods Museum in Beijing on 14th September when the artist himself provided a guided tour and explained the connotations of the character '*Huang*' (恍) for the exhibition title as 'suddenly', 'seemingly', and 'ecstasy'. I then conducted a two-hour interview with Liang on the 17th at the office building of MWoods, after which I realised that I need to brush up on my knowledge of silkworm biology to better understand Liang's artistic experiments. The technical skills required to realise his silkworm projects were still too vague and lacked tangible details by merely talking with the artist and reading art critics' articles. I was determined to narrow my doctoral research to Liang's works only and turned to a more anthropological approach. Meanwhile, in the following course of my learning mulberry-silkworm biological knowledge from textual materials, I happened to notice a thriving silkworm husbandry industry in my hometown Hechi City which I had ignored in my adolescence and later needed to refamiliarise and revisit for this project.

At the end of March in 2019, I met Liang Shaoji again in Wuzhen County of Zhejiang Province, where the second Wuzhen Contemporary Art Exhibition 'Now is the time' was held and three of Liang's artworks were among the group exhibition there. I attended several talks by artists and curators including Liang. I also visited an old silk reeling factory site situated at the heart of a designated tourist area where workers were showing visitors how to boil cocoons and reel silk on a treadle machine. I interviewed these workers briefly in order to learn about the current situation in this former 'silkworm province'. Although silkworm husbandry in Zhejiang is not within my research scope, this fragmented information helped me to better understand the general picture of sericultural development in China. Another group exhibition 'Growing' I visited was held at Chronus Art Centre in Shanghai where Liang's works were juxtaposed with other world-renowned bio-artists such as Eduardo Kac and the Tissue Culture & Art Project. On-site experiencing artworks is surely a good channel to engage with and comprehend them, but I am more interested in the backstage stories between the artist and the silkworms, so as to make a more accurate interpretation supported by details of his art-making processes.

Besides three gallery visits, I also went to Liang Shaoji's art studio and home located in Tiantai County, Zhejiang Province in early April 2019. Liang rents a single office room as his home within Tiantai Museum, a complex designed by architect Wang Shu. He didn't cook himself but dined in the staff canteen with the museum's security guards; he used the building's communal facilities such as bathrooms and toilets to keep a minimalist life. His room was converted from an office, furnished only with a bed and desk, and abundant books, magazines, journals, and drawing drafts piled from floor to ceiling. Liang rented another large office in the museum complex and a car park on the opposite of the street as his two studios (Figure 1.6). During my stay in Tiantai, I saw a large number of unfinished or unexhibited works and sericultural tools in the artist's studio; I met his local friends and other museum staff to have dinners and chats. Liang told me that at one point he was conducting the whole process of raising

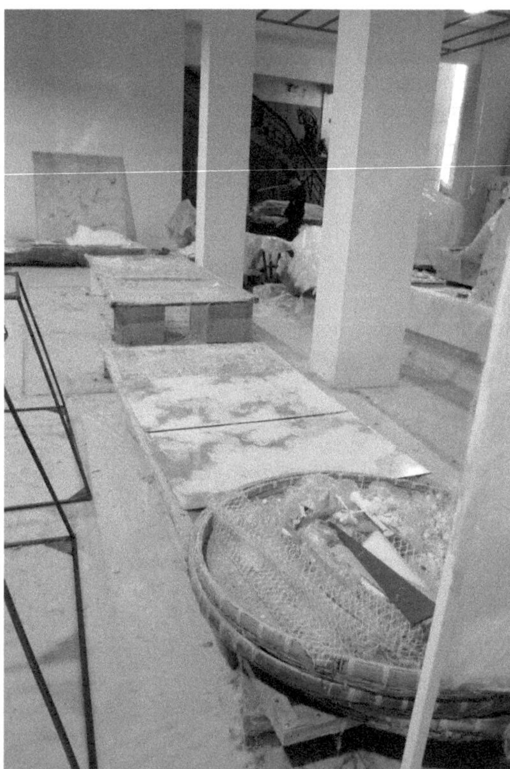

Figure 1.6 Liang's studios in Tiantai, Zhejiang. Photo by author

silkworms and creating his artworks in his studio in Tiantai, but for one thing, the bad smell of diseased silkworms was too strong for museum visitors, and the cost of transporting fresh mulberry leaves from mulberry fields to the downtown area daily was prohibitive. He thus rented a plot of land in a rural village in the neighbouring county Linhai to plant mulberry and feed silkworms. Until the larva reached their fifth instar before the cocooning phase, they were moved by car to Liang's studio in Tiantai to join the artwork creation.

Besides, Liang's studio in Tiantai sits at the foot of Tiantai Mountain by the entrance to a national scenic site. The ancestral temple of the Buddhist Tiantai sect, Guoqing Temple, is located on this hill. Liang conducted a performative art project named 'Cloud Mirror' in 2007 where he sat at the peak of Tiantai Mountain with a mirror on his knee and live silkworms were spinning silk filaments on the flat mirror with the clouds' reflection. Climbing Mount Tiantai and visiting Guoqing Temple gave me a more tangible understanding of both the 'Cloud Mirror' performance left only on video, and the local Buddhist-Daoist culture of Tiantai, which Liang claimed deeply influenced his artworks.

I had extensive conversations with Liang in Tiantai, although the April climate and temperatures in Zhejiang were still too chill for sericultural practice and therefore local silkworm season is a month later than that in subtropical Guangxi. I hence did not witness how Liang Shaoji took care of silkworms in Linhai or Tiantai, but the visit to his studio is still fruitful where I got to know the artist's living and working environment, his acquaintances, and the oral history of his whole life journey.

As a supplement for not being able to actually participate and observe how the artist raises silkworms and creates artworks on-site in Zhejiang Province, I returned to Hechi of Guangxi where I can more conveniently live with silkworm farmers for a longer period of time to gain first-hand practical knowledge about this insect and how to take care of them. Yizhou County in Hechi City became the largest cocoon and raw silk production centre nationwide and globally in the recent decade, due to the industry transformation policy 'East Mulberries Moving West' officially implemented by the Chinese government in 2006. Thereby, less economically developed southwestern provinces with lower labour costs, namely, Guangxi, Yunnan, and Sichuan gained the state support out of this initiative and gradually replaced the historical 'silkworm regions' such as Guangdong at Pearl River Delta and Zhejiang and Jiangsu at Yangtze River Delta where silkworm husbandry is shrinking, and more governmental investment instead is devoted to the silk processing industry. I grew up in the downtown area of Hechi until I left home in 2010 to attend university in Beijing. However, it was not until I started working on this doctoral research that I had any knowledge of Yizhou County seventy kilometres away as an emerging silkworm husbandry centre in China and the global world.

My identity as a local has largely facilitated my fieldwork. On the one hand, I was able to find desired gatekeepers more quickly through my family's social network of acquaintance in Hechi (known as 'relations (*guanxi*)' in Chinese social codes) and gain access to levels of government departments, silk reeling and processing enterprises, research institutes and sericulturists in the whole industry. In early April when the silkworm season of 2019 began, I rented a bedroom in a farmer's farmhouse in Village Shangping in the northern suburb of Yizhou County to begin my long-term research in this rural area. As I speak the local dialect, my daily communication with people of all occupations and ages encountered no barriers. I was involved in the daily labour of raising silkworms for the family where I resided, to obtain my own embodied and sensory knowledge of nurturing silkworms. I also spent time wandering around the lanes and mulberry fields, chatting with local residents and joining in occasional gatherings within the village. Most of the villagers raise silkworms at home; some are too old or physically challenged to do the heavy work of picking leaves, but they would assist the feeding or at least know something about silkworms. I have a distant relative who lives in another village on the western outskirts of Yizhou, half an hour's drive from Shangping. I also visited that village three times for some possible similar or different voices. My months' experience of cohabiting with silkworms and farmers in Shangping guarantees me rich insights into the silkworm as a companion species, raising silkworm as a livelihood, and sericulture as an industry, which allowed me to attend to sharp details and ask pertinent questions when interacting with silkworm scientists and the artist Liang.

Besides living, visiting, and learning with silkworm farmers in villages on the outskirts of Yizhou, I also made several day trips to the town centre of Yizhou and visited the government's sericultural office, the workshops of silk reeling and pupa processing factories and a state-run silkworm breeding station. I obtained the last three years' industry reports of sericulture in Yizhou from the sericultural office, from which I gained a panoramic view of silkworm husbandry in this region. In my frequent visits to the silkworm breeding station, I interviewed the chief scientist as a manager and several female breeders there, learned how to cut the cocoons to get live pupae for mating and reproduction from these women, and observed the entire process of silkworm egg production. This gave me an intuitive glimpse into an alternative fate of cocoons sold by farmers other than being boiled on the reeling machines, which can be further compared with the fates of silkworms raised and employed for the artistic mission by Liang Shaoji.

Moreover, as Liang's artistic techniques to work with silkworms draw heavily on the outcomes of scientific studies of sericulture and hands-on support from biologists on crossbreeding, I approached silkworm scientists to deepen my comprehension of a more pragmatic experiment-based method to study animals. I went to the capital city of Guangxi Province, Nanning, for short stays in March, April, and November, where the provincial research institute Guangxi Sericulture Research Centre is located. I chose this institute for my focused research on silkworm biology because it is a branch of

Guangxi Academy of Agricultural Sciences. My late grandmother used to work at the academy as a crop scientist and her former colleagues kindly acted as my gatekeepers, so that I was allowed to carry out short-term archival and anthropological research at the labs and the library in the sericulture institute. I mainly conducted participant observation when staying in the biologists' labs, noting them making samples, recording results, taking photos, and discussing the experiment processes. I was merely not allowed to enter the lab working on breeding for resistance to pebrine disease, as I was not familiar with the code of conduct in such a lab environment and possible leakage of this pathogen would lead to negative consequences. In addition to casual talks with present scientists, I conducted two in-depth semi-structured interviews with a chief biologist Mr Zhang at this institute, who kindly answered my additional questions on the instant messaging app in the later stage of my research after the fieldwork. Equipped with more knowledge on the scientific epistemology towards silkworms, I was able to critically examine the artist's ongoing cross-disciplinary collaborations with silkworm biologists over the last 30 years.

In my multi-sited fieldwork, I carried a notebook and pen most of the time to take notes and draw rough sketches. Only planned interviews with the interviewees' oral permission were digitally recorded by my voice recorder. The interviews were conducted in Hechi dialect or Mandarin, and I made factual transcriptions by myself and translated sections used in this book into English. My field diary contained a record of my everyday witness, immediate emotional reactions, and thoughts written at night from memory. In addition, I also refer to archives from Guangxi Sericulture Research Centre, government work reports and industry annual reports from the sericulture office of Yizhou County, media reports, art critiques, historical records, academic papers on sericultural science, and online chats with my interlocutors. I tagged these textual materials with keywords, which allowed me to better identify commonalities in silkworm-raising experiences and techniques across arts, science, and agriculture. Besides, I carried a portable camera for photography and video recording; these visual and audio materials serve both for my analysis and possible multi-media sharing of my research outcome other than a book later. Due to the limited mobility in COVID-19 pandemic since 2020, I was unable to return to mainland China to conduct follow-up fieldwork research as much and therefore my comprehension of the impact of COVID-19 on the industry as a whole and individuals involved is limited and incomplete; for this reason, in this book, the situations in and after 2020 will not be focused.

Multispecies Ethnography

To reflect on my research methodologies regarding the collaborative relationship between silkworms and artist Liang Shaoji, I begin by reviewing early anthropological approaches to animals. Anthropology has its long tradition of exploring other living organisms 'as part of the landscape, as food for humans, [and] as symbols' (Kirksey and Helmreich 2010: 545) in its archaic discourses of totemism, savage animism, and primitivism. Animals and plants in nature, being the familiar targets of anthropological inquiries, have served as mirrors and windows for analysis of human societies and cultures (Mullin 1999), where abundant ethnographic accounts attempt to delineate local cosmologies, ecological knowledge, and dwelling strategies in many non-Western, pre-industrial communities. Domesticated species, in particular, are never absent from the disciplinary lineage of anthropology, especially in studies of agrarian and pastoral societies. Animals are prominent in the early anthropological literature on evolutionary patterns of societies (Morgan 1868), social organisations (Mauss 1967), and structuralist totemism (Lévi-Strauss 1966). Although these theoretical frameworks on human society and culture are accused of instrumentally presenting and using animals and thus seem human-centric, Galvin defends these studies' attention and recognition of mutual dependencies, nonhuman agencies, and entanglements of human and non-human lives (2018: 235). Ingold also regards the current 'animal turn' in social sciences and humanities as nothing significantly new for anthropology and plants and animals are neither inert nor passive but usually 'ontologically equivalent' in manifold places, periods, and communities in anthropological inquiries (1980; 2013). For those familiar with ecological studies of hunter-gatherer societies, pastoralism or farming, as Ingold claims, 'the oft-repeated claim of material culture theorists – namely, that the "non-human" has been marginalised or suppressed in the social sciences – seems preposterous' (2013: 16).

ME, however, emerged within this discipline in the recent decade to discover and learn more egalitarian possibilities in contingent multispecies relationships in more contemporary situations. It encourages exploration into the natural-cultural borderlands and concentration on the entanglements of agentive matters and organisms in a multiplying assemblage (Kirksey and Helmreich 2010). This approach can be associated with the Deleuzian concept of 'becoming', both of which abound in 'non-hierarchical alliances, symbiotic attachments and the mingling of creative agents' (Kirksey and Helmreich 2010: 458). Tsing's ME, as an exemplary case study, delves into the commercial mushroom collected from the seams of the global capitalist market, where the mushroom is portrayed as a symbiotic companion species often flourishing at the marginal zones such as the ruins at Chernobyl despite the severe ecological destruction (2012: 152). The research method Tsing proposes is to visit the field enough to get familiar with the species' assemblages and body forms in a certain landscape as their expression of sociality (2013: 32). She also contends that 'critical description' should be employed in ME studies on more-than-human sociality, which requires a sensibility to appreciate and learn with other organisms through common activities (2013). Research focuses on micropolitics and microbiology featured in Tsing's mushroom-centred ethnography are too prominent in Paxson's study on microbial entanglements in pasteurisation and raw-milk cheese production. Focusing on the food value and ethics attached to the Pasteurian standards in the cheese industry, Paxson (2008) examines issues of food safety, state regulation, and customer culture around cheese consumption.

Ethics is another crucial aspect for consideration in ME literature. What is conventionally regarded polite and nice in interpersonal communication would not be necessarily valid and plausible in trans-species interaction and therefore cannot be simply imposed as measuring standards into new situations. Candea (2010), for instance, takes a close observation of biologist-meerkat relations at a research station in South Africa and advocates the productive values of cultivated detachment instead of commonly praised connection and engagement in human-animal relations. As Candea notes, detachment, though often passively associated with moral inferiority, uncaringness, cold-heartedness, and indifference, can enable mutual suspension of action and the maintenance of a polite distance. He, therefore, proposes an ethical stance of inter-patience where 'action and inaction, engagement and detachment are inextricably entwined', allowing 'the extension of the meaning of relationship beyond intersubjectivity and interaction' (2010: 249). Blanchette also provides an astonishing ethnographic account of sows 'that cannot be pet' in farm factories (2019). The unspoken terms in industrial animal husbandry say that unexpected petting and caressing could surprise the pregnant sows and cause them to startle, writhe, or even abort in the cage, as Blanchette introduces, and one sow's miscarriage out of shock could even pass down like a wave to a row of its peers nearby (2019). Those pigs are weakened by uniformly working routines and therefore cannot afford 'excess affect' from human workers or visitors in the workshop (2019: 69).

From a more ontological perspective, Kohn (2013) in his study of dogs and their Runa masters' in Amazonian forests probes into indigenous cosmology and cross-species communication in their metaphoric dream interpretation. People there believe that through dreaming, consuming hallucinogens and tobacco they can gain access to the perspectives of other beings such as dogs, but entering canine subjectivity risks the loss of human beings' privileged ontological status (Kohn 2013: 143). Acknowledging nonhuman ontologies and capacities to make worlds, the Amazonia style of perceiving nature and engaging with the biosphere delineated by Kohn is an exemplary presentation of 'perspectivism' in ME (also in de Castro 2004). Kohn's insightful perspectivist approach to nonhumans is praised by Ingold as nonhumans in such cosmological framework can communicate without a human spokesperson on their behalf 'simply by virtue of their presence and activity' (2013: 20). Descola instead criticises the limits of Kohn's works as unable to be applied in other types of ecosystems, because each lake, forest and hill has its distinct networks, life forms, landscape features and therefore has a unique way of 'thinking and speaking' (2014: 270). Descola then points to the direction of further research along with the genealogy of ME to explore 'how reciprocal interpretations of behavioural and environmental signs have built up the respective knowledge of coevolving humans and animals' (2014: 272).

More severe criticism of ontological turn in ME comes from Graeber with his perceptive awareness of power relations and politics. He condemns the excessive dedication to understanding and describing forms of 'radical alterity' as too essentialist, de-historicising and conservative regarding its potential result of 'protect[ing] more structurally powerful and authoritative ontological positions from [any] challenge' (2015: 7). Another disapproving voice against ME as a research method targets its grounding concept 'species'. Haraway explicitly rebukes the term as unsteady and oxymoronic to indicate biological distinction and familiarities (2008). Ingold similarly denounces it as a sovereign perspective that places humanity as a central and universal referential point to recognise biodiversity or species multiplicity (2013: 19). When the logic of multispecies is to associate individuals by their received likeness and to divide them by diversification, Ingold on the contrary contends that difference connects and similarity divides (2013: 20). In his proposed project 'anthropology beyond humanity', anthropologists distinguish themselves from other disciplines not by the research objects but by the unique way of working and learning through participation in other lives. He revives the idea of 'craft' from Mills (1956 in Ingold 2013) and argues that there is no division between method and theory as well as work and life (2013: 82). Anthropology hence is the craftsmanship of 'studying with' rather than 'studying of' (2013). Despite all the disputes on the terminology for analysing interspecies symbiosis, the above literature in both theoretical and methodological level paves new directions for further exploration in human-animal interaction and communication, from which I draw inspiration for my own research on the artist-silkworm sociality.

A posthuman take on domestic animals such as the silkworm is to include them into the sociality with humans, which used to be conspecific within interpersonal connections and now is expanded to interspecies interactions (Haraway 2008; Latour 2004). Compared to wild animals, those domesticated ones are thought to have more 'experience' in socialising with human beings and therefore it is more likely to detect their attitudes towards us as close acquaintances (Coy 1989). Following ME's proposal to (re-)imagine the 'webs of interspecies dependencies' (e.g. Tsing 2015), this book compared Liang's artistic co-working with domestic silkworms with other forms of human-silkworm entanglements in agriculture and science in China, to more accurately capture the uniqueness of his artistic techniques and conceptual adaptability of '*ziran*' along with the trans-species 'collaboration'.

As a domesticated species, silkworms cohabit with farmers in rural China village houses, where cocoon production sustains many families.

Liang Shaoji also nurtured silkworms in Linhai for years, suggesting physical, if not affective, intimacy. As a tactic approach to study silkworms, I engaged their human caretakers as interlocutors and 'study with' them (cf. Ingold 2013), fully aware of critiques that ME might extend anthropological 'otherness' to nonhumans (DiNovelli-Lang 2013). Meanwhile, I apprenticed in sericulture, learning the craft of raising silkworms as an amateur to gain embodied knowledge as a field method (Coy 1989). Pragmatic learning provided access to both verbal discourse and the non-verbal, kinaesthetic sharing of sericultural techniques and care practices (Cohen 2019). Regular 'toil', coupled with reflective engagement with mistakes and progress (Cohen 2019; Marchand 2010), offered ethnographic insights into the ethics of politeness and manipulation, strategies of multispecies collaboration – harmonious or exploitative – and the cosmological dimensions of ecology in Liang and the silkworms' co-making of art.

References

Ames, R.T., 2001. The Local and the Focal in Realizing a Daoist World. *In*: N.J. Girardot, J. Miller, and X. Liu, eds. *Daoism and Ecology: Ways within a Cosmic Landscape*. Cambridge: Harvard University Press, 265–82.

Ames, R.T. and Hall, D.L., 2003. *Daodejing 'Making This Life Significant': A Philosophical Translation*. New York: Ballantine Books.

Baker, S., 2000. *The Postmodern Animal*. London: Reaktion Books.

Baker, S., 2013. *Artist/Animal*. Minneapolis: University of Minnesota Press.

Barad, K., 2003. Posthumanist Performativity: Toward an Understanding of How Matter Comes to Matter. *Signs*, 28 (3), 801–831.

Bennett, J., 2010. *Vibrant Matter: A Political Ecology of Things*. Durham: Duke University Press.

Bhabha, H., 2002. *The Location of Culture*. New York: Routledge.

Blanchette, A., 2019. Making Monotony: Bedsores and Other Signs of an Overworked Hog. *In*: Sarah Besky and A. Blanchette, eds. *How Nature Works: Rethinking Labor on a Troubled Plane*. Albuquerque: University of New Mexico Press, 59–77.

Bowden, R., 2004. A Critique of Alfred Gell on Art and Agency. *Oceania*, 74 (4), 309–324.

Braidotti, R. and Hlavajova, M., 2018. *Posthuman Glossary*. London: Bloomsbury Academic.

Brouwer, M., 2007. Cloud (on Liang Shaoji's Work). *Yishu: Journal of Contemporary Chinese Art*, 6 (3), 75–79.

Candea, M., 2010. 'I Fell in Love with Carlos the Meerkat': Engagement and Detachment in Human–Animal Relations. *American Ethnologist*, 37 (2), 241–258.

Cheng, C., 1976. A Model of Causality in Chinese Philosophy. *Philosophy East and West*, 26, 12–18.

Cheng, M., 2007. Animalworks in China. *TDR: The Drama Review*, 51 (1), 63–91.

Cheng, M., 2015. 蠶語/人易: 梁紹基的'元' [online]. Available from: https://www.academia.edu/7952556/_%E8%A0%B6%E8%AA%9E_%E4%BA%BA%E6%98%93_%E6%A2%81%E7%B4%B9%E5%9F%BA%E7%9A%84_%E5%85%83_.

Ching, J., 1993. *Chinese Religions*. London: MacMillan.
Chu, K., 2015. Animating *Shanshui*: Chinese Landscapes in Animated Film, Art and Performance. *In*: C. Pallant, ed. *Animated Landscapes: History, Form, and Function*. London: Bloomsbury Academic, 109–124.
Cohen, E., 2019. Anthropology of Knowledge. *Journal of the Royal Anthropological Institute*, 16, 193–202.
Coy, M.W., 1989. *Apprenticeship: From Theory to Method and Back Again*. Albany: SUNY Press.
Da Costa, B. and Philip, K., 2008. *Tactical Biopolitics: Art, Activism, and Technoscience*. Cambridge: MIT Press.
de Castro, V., 2004. Exchanging Perspectives: The Transformation of Objects into Subjects in Amerindian Ontologies. *Common Knowledge*, 10 (3), 463–484.
Demos, T.J., 2016. *Decolonizing Nature: Contemporary Art and the Politics of Ecology*. Berlin: Sternberg Press.
Descola, P., 2014. All Too Human (Still): A Comment on Eduardo Kohn's How Forests Think. *HAU: Journal of Ethnographic Theory*, 4 (2), 267.
DiNovelli-Lang, D., 2013. The Return of the Animal: Posthumanism, Indigeneity, and Anthropology. *Environment and Society*, 4, 137–156.
Eisenman, S.F., 2013. *The Cry of Nature: Art and the Making of Animal Rights*. Islington: Reaktion Books.
Firth, R., 1992. Art and Anthropology. *In*: J. Coote and A. Shelton, eds. *Anthropology, Art, and Aesthetics*. Oxford: Oxford University Press, 15–39.
Fok, S., 2012. *Life and Death: Art and the Body in Contemporary China*. Bristol: Intellect.
Fortis, P., 2010. The Birth of Design: A Kuna Theory of Body and Personhood. *Journal of the Royal Anthropological Institute*, 16 (3), 480–495.
Galvin, S.S., 2018. Interspecies Relations and Agrarian Worlds. *Annual Review of Anthropology*, 47, 233–249.
Gell, A., 1992. The Technology of Enchantment and the Enchantment of Technology. *In*: J. Coote and A. Shelton, eds. *Anthropology, Art, and Aesthetics*. Oxford: Oxford University Press, 40–63.
Gell, A., 1998. *Art and Agency*. Oxford: Clarendon Press.
Gladston, P., 2011. A Conversation with Liang Shaoji. *In*: *Contemporary Art in Shanghai: Conversations with Seven Chinese Artists*. Hong Kong: Blue Kingfisher.
Gow, P., 1999. Piro Designs: Painting as Meaningful Action in an Amazonian Lived World. *Journal of the Royal Anthropological Institute*, 5 (2), 229–246.
Graeber, D., 2015. Radical Alterity is Just Another Way of Saying 'Reality': A Reply to Eduardo Viveiros de Castro. *HAU: Journal of Ethnographic Theory*, 5 (2), 1–41.
Haraway, D.J., 1991. *Simians, Cyborgs, and Women: The Reinvention of Nature*. New York: Routledge.
Haraway, D.J., 2008. *When Species Meet*. Minneapolis: University of Minnesota Press.
Ingold, T., 1980. *Hunters, Pastoralists and Ranchers: Reindeer Economies and their Transformations*. Cambridge: Cambridge University Press.
Ingold, T., 2013. *Making: Anthropology, Archaeology, Art and Architecture*. London: Routledge.
Isto, R., 2015. Organic (Un)ground in the Time of Biopower and Hyperobjects: Conceptualizing Global Posthumanism in the Art of Xu Bing and Gu Wenda. *Journal of Contemporary Chinese Art*, 2 (2–3), 195–215.

Jenkins, T.N., 2002. Chinese Traditional Thought and Practice: Lessons for an Ecological Economics Worldview. *Ecological Economics*, 40 (1), 39–52.
Jevbratt, L., 2009. Interspecies Collaboration – Making Art Together with Nonhuman Animals [online]. Available from: http://jevbratt.com/writing/jevbratt_interspecies_collaboration.pdf.
Kirksey, E., ed., 2014. *The Multispecies Salon*. Durham: Duke University Press.
Kirksey, S.E. and Helmreich, S., 2010. The Emergence of Multispecies Ethnography. *Cultural Anthropology*, 25 (4), 545–576.
Kisin, E. and Myers, F.R., 2019. The Anthropology of Art, After the End of Art: Contesting the Art-Culture System. *Annual Review of Anthropology*, 48 (3), 317–334.
Kohn, E., 2013. *How Forests Think: Toward an Anthropology Beyond the Human*. Berkeley: University of California Press.
Latour, B., 1999. *Pandora's Hope: Essays on the Reality of Science Studies*. Cambridge: Harvard University Press.
Latour, B., 2004. *Politics of Nature: How to Bring the Sciences into Democracy*. Cambridge: Harvard University Press.
Latour, B., 2017. *Facing Gaia: A New Enquiry into Natural Religion*. Cambridge: Polity Press.
Layton, R., 1991. *The Anthropology of Art*. Cambridge: Cambridge University Press.
Leach, E., 1973. Levels of Communication and Problems of Taboo in the Appreciation of Primitive Art. *In*: A. Forge, ed. *Primitive Art & Society*. Cambridge: Cambridge University Press.
Lévi-Strauss, C., 1966. *The Savage Mind*. Paris: Librairie Plon.
Liang, S., 1994. About the Nature Series [online]. *ShanghART Gallery*. Available from: https://www.shanghartgallery.com/galleryarchive/texts/id/54.
Liang, S., 2002. About the Nature Series: Notes on Creative Work. *In*: W. Ai, ed. *Chinese Artists, Texts and Interviews: Chinese Contemporary Art Awards*. Hong Kong: Timezone 8, 41–46.
Liang, S., 2008. Notes on the Creation of 'Broken Landscape' [online]. *ShanghART Gallery*. Available from: https://www.shanghartgallery.com/galleryarchive/texts/id/10010.
Liang, S., 2014. *'Back to Origin' Catalogue*. Shanghai: ShanghART.
Liu, A., 2011. The Living Word: Xu Bing and the Art of Chan Wordplay. *In*: H. Tsao and R.T. Ames, eds. *Xu Bing and Contemporary Chinese Art: Cultural and Philosophical Reflections*. Albany: SUNY Press, 117–146.
Marchand, T.H., 2010. Making Knowledge: Explorations of the Indissoluble Relation between Mind, Body and Environment. *Journal of the Royal Anthropological Institute*, 16, 1–21.
Marcus, G.E. and Myers, F.R., eds., 1995. *The Traffic in Culture: Refiguring Art and Anthropology*. Berkeley: University of California Press.
Mauss, M., 1967. *The Gift: The Form and Reason for Exchange in Archaic Societies*. New York: Norton.
Miller, D., 1985. *Artefacts as Categories*. Cambridge: Cambridge University Press.
Morgan, L.H., 1868. *Systems of Consanguinity and Affinity of the Human Family*. Washington: Smithsonian Institution.
Morphy, H., 2009. Art as a Mode of Action: Some Problems with Gell's Art and Agency. *Journal of Material Culture*, 14 (1), 5–27.
Morphy, H., 2013. The Anthropology of Art. *In*: T. Ingold, ed. *Companion Encyclopedia of Anthropology*. London: Routledge, 648–685.

Morton, T., 2007. *Ecology without Nature: Rethinking Environmental Aesthetics*. Cambridge: Harvard University Press.

Mullin, M.H., 1999. Mirrors and Windows: Sociocultural Studies of Human-Animal Relationships. *Annual Review of Anthropology*, 28, 201–224.

Munn, N., 1977. The Spatiotemporal Transformations of Gawa Canoes. *Journal de la Société des Océanistes*, 33, 39–53.

Needham, J., 1956. *Science and Civilisation in China Vol. II: History of Scientific Thought*. Cambridge: Cambridge University Press.

Paxson, H., 2008. Post-Pasteurian Cultures: The Microbiopolitics of Raw-Milk Cheese in the United States. *Cultural Anthropology*, 23 (1), 15–47.

Prince Claus Fund, 2009. *Culture & Nature: Prince Claus Awards 2009*. Amsterdam: Prince Claus Fund.

Schneider, A., ed., 2017. *Alternative Art and Anthropology: Global Encounters*. London: Bloomsbury.

Steiner, C.B., 1994. *African Art in Transit*. Cambridge: Cambridge University Press.

Stracey, F., 2009. Bio-art: The Ethics Behind the Aesthetics. *Nature Reviews Molecular Cell Biology*, 10 (7), 496–500.

Strathern, M., 1980. No Nature, No Culture: The Hagen Case. *In*: C.P. MacCormack and M. Strathern, eds. *Nature, Culture and Gender: A Critique*. Cambridge: Cambridge University Press.

Tsao, H., 2011. Reading and Misreading: Double Entendre in Locally Oriented Logos. *In*: H. Tsao and R.T. Ames, eds. *Xu Bing and Contemporary Chinese Art: Cultural and Philosophical Reflections*. Albany: SUNY Press, 1–32.

Tsao, H. and Ames, R.T., eds., 2011. *Xu Bing and Contemporary Chinese Art: Cultural and Philosophical Reflection*. Albany: SUNY Press.

Tsing, A.L., 2013. More-than-Human Sociality: A Call for Critical Description. *In*: K. Hastrup, ed. *Anthropology and Nature*. London: Routledge, 27–42.

Tsing, A.L., 2015. *The Mushroom at the End of the World: On the Possibility of Life in Capitalist Ruins*. Princeton: Princeton University Press.

Tu, W.-M., 1989. The Continuity of Being: Chinese Visions of Nature. *In*: J.B. Callicott and R.T. Ames, eds. *Nature in Asian Traditions of Thought: Essays in Environmental Philosophy*. Albany: University of New York Press, 67–78.

Turner, V., 1967. *The Forest of Symbols*. Ithaca: Cornell University Press.

Vinograd, R., 2011. Making Natural Languages in Contemporary Chinese Art. *In*: H. Tsao and R.T. Ames, eds. *Xu Bing and Contemporary Chinese Art: Cultural and Philosophical Reflections*. Albany: SUNY Press, 95–116.

von Drathen, D., 2011. Liang Shaoji: Of the Immaterial and Its Corporeality. *In*: M. Bloemheuvel and T. van Kooten, eds. *Windflower: Perceptions of Nature*. Rotterdam: NAi Publishers.

Weintraub, L., 2012. *To Life!: Eco Art in Pursuit of a Sustainable Planet*. Ewing: University of California Press.

Xia, K., 2015. The Silk Shadow of Chora-topia [online]. *ShanghART Gallery*. Available from: https://www.shanghartgallery.com/galleryarchive/texts/id/8092.

Xia, K., 2017. Kiefer and Liang Shaoji: Differences of Modernity between East and West in Integrating Three Worlds. *Northern Art*, 12, 60–61.

Yang, J., 2017. Pursuit for the Tao through the Silkworm: A Conversation with Liang Shaoji [online]. Available from: https://jyx.jyu.fi/bitstream/handle/123456789/59988/Liang%20Shaoji.pdf?sequence=1.

Yang, J., 2018. Dao of the Silkworm: The Ecological Thrust of Liang Shaoji's Nature Series. *New Arts: Journal of the National Academy of Art*, 39 (6), 119–127.
Zheng, B. and Lee, S., 2016. Contemporary Art and Ecology in East Asia. *Journal of Contemporary Chinese Art*, 3 (3), 215–222.
Zulueta, C.C., 2016. Two Artists, Two Ecologies, and a Shared Empathy Towards Non-Human Animals' Agencies: Yanagi Yukinori and Liang Shaoji. *Journal of Contemporary Chinese Art*, 3 (3), 377–387.

2 The Cultivator and Maker

In January 2022, China's urban landscape was still subdued by governmental pandemic restrictions that rendered inter-city travel a complex endeavour. Perhaps this contextual constraint explained the curiously modest attendance at the large-scale retrospective exhibition 'Liang Shaoji: A Silky Entanglement' at Power Station of Art, one of Shanghai's most distinguished contemporary art institutions. On the tenth day of that month, I traversed from Guangxi to Shanghai, having concluded my doctoral dissertation defence online a few weeks ago. A temporal interlude of freedom beckoned, though spatial mobility remained stringently circumscribed by the ubiquitous COVID-related QR codes, those digital sentinels of pandemic control.

Within the two-story gallery, fewer than ten visitors wandered, creating an atmosphere at once unprecedentedly intimate to each artwork though with lingering poignancy. In the dim room displaying the installation 'Listening to the Silkworm' (Figure 2.1), a solitary staff stood in a shadowy corner while a single viewer, seated on a cushion, gazed at her mobile phone. Behind the translucent curtain, the anticipated soundscape – live recordings of silkworms chewing mulberry leaves, intended to be transmitted through strategically placed headphones – was eerily absent. Only a few white cocoons nestled sparsely among dried stalks remained as silent components on the shelves.

'Silkworm cultivation already ceased in December'. The gallery attendant murmured, with pity in his voice. 'Too cold. You could still have seen some silkworms here in October and November'. When I inquired whether he had personally tended to the living silkworms in preceding months, he hesitated, explaining that specialised caretakers were responsible for feeding, clearing their excretion, and the delicate process of segregating sick worms emitting aberrant odours.

This artwork, originally conceptualised as a profound auditory experience, now resided quietly, still subtly inviting a meditation on absence, sensation, and ephemerality. Such a quietude of 'Listening to the Silkworm' resonated with my memories of a conversation I had once shared with Liang Shaoji in his studio nestled at the foot of Tiantai Mountain, emerging from a seemingly naive and random inquiry. Little did I realise then how profoundly this brief

DOI: 10.4324/9781003569091-2

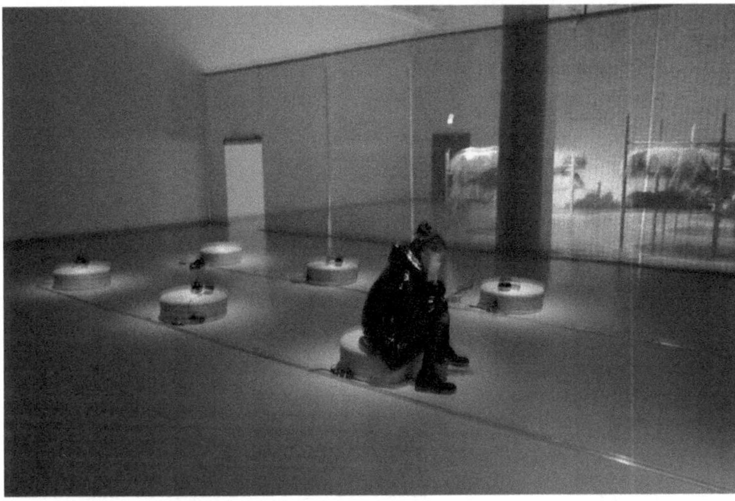

Figure 2.1 'Listening to the Silkworm/Nature Series No. 96', 2022. Photo by author

chat would later unravel a rich tapestry of silkworm cultivation, where the boundaries between human experience and the sensory world of non-human beings become wonderfully, almost mystically, permeable.

'Do Silkworms Enjoy Music?'

Liang didn't answer my question directly. Instead, he shared a piece of his experience in rural Linhai around 20 years ago in the 1990s:

> At that time, I lived in the countryside. Local villagers were often listening to Yue Opera on cassette tapes when feeding the silkworm larvae in their farmhouses. I once asked them: "Do silkworms also enjoy listening to Yue opera?" One farmer Liu replied, jokingly: "We delight in it, so do silkworms (我们欢乐, 蚕也欢乐)." Other villagers all laughed and agreed.

Yue opera, also known as Shaoxing opera, has been a popular Chinese operatic genre in Zhejiang province since the early twentieth century. Besides occasional live performances, the most accessible ways around the year 2000 to enjoy Yue Opera for people in rural areas in Zhejiang were cassette tapes and music programmes on FM broadcasts.

'We delight in it, so do silkworms' seemed to express how silkworms in Linhai can not only 'hear' but also acquire a similar appreciation of Yue opera as their human raisers do. The lively gaiety in the space was brought by Yue opera melodies, which were usually played on affordable devices such as tape

recorders or the radio for local villages. This seemingly imposed local taste of music, as assumed by Linhai farmers, appears to naturally traverse the species boundaries and be shared by all creatures inhabiting the same village. Despite heterogeneous sensing apparatus, the villagers still made a jokingly yet wishful presumption that silkworms can have fun in Yue opera, the same as humans do during their daily interactions in cultivating labour.

Silkworm Cultivation in China

Liang Shaoji has acquired not only practical knowledge and skills in raising silkworms but also artistic inspirations from everyday toil in the farm, as exemplified in 'Listening to the Silkworm' which I will return to discuss later in this chapter. But to contextualise the first decade of Liang's artistic career in the 1990s when he lived and worked with silkworm farmers to familiarise himself with his animal collaborators, I first sketch the domestication of silkworm and the labour-intensive silkworm husbandry in China. The timeline of silkworm domestication (*Bombyx mori*) stretches over more than 3000 years. The silkworm's relationship to human society has ranged from silk manufacture and consumption to artistic expression, agricultural production and, in modern times, bioscientific experiments. Silk has been used as clothing, but also as a medium of exchange, a writing and painting surface as well as diplomatic offerings in ancient China from around 3000 BC, thus forming a quintessential part of Chinese material culture. In contemporary society, both silk and the silkworm have been commoditised as organic bio-product in local and global markets, boosted by fast-developing biotechnologies such as transgenesis, transplantation, and gene editing. Exquisite silk garments and silk fabrics as painting surfaces have been the most common material channels in which silkworms are indirectly associated with the realm of art. In contemporary art worlds, the way silkworms engage in art creation takes diverse forms, such as in Liang Shaoji's '*ziran* series' where those low-profile off-stage creatures have been cultivated to confront more challenging artistic tasks.

The English term 'sericulture' is a teleological compound. In 'silk + cultivation', the end-product silk has been placed at the centre of the whole industry. However, raw silk is not 'cultivated' but comes from filature where cocoons are reeled into threads usually in boiling water. What farmers cultivate in sericultural farming are silkworms and mulberry trees, two species that are prominent in the Chinese counterpart of the term sericulture, '*sang-can*' (桑蚕), literally 'mulberry and silkworm'. They have accompanied each other throughout the history of sericultural practice and knowledge production. In silkworm husbandry for large-scale silk industry in China, fresh mulberry leaves remain the sole nutrition and energy source for this species whose survival and reproduction are highly dependent on its human raisers (Zhao 2005). The labour-intensiveness of silkworm farming lies in the astonishing growth of silkworm caterpillars, whose body size can multiply by a factor of 10,000 within its three weeks' larval phase after hatching from ant-size eggs (Field 2014). The lifespan of a domestic silkworm normally lasts around 50 days in total, and farmers are only involved in half of their lifetime from newly hatched larvae in the first instar (i.e. a phase between two periods of moulting) to pupae in freshly spun cocoons. In 20–30 days, silkworms go through dramatic morphological changes and their human keepers have to meet their physiological needs in a very compressed timeline.

In oral and written practices of sericulture in China, the unit of quantity for silkworm is '*zhang*' (张), literally a sheet, which refers to the number of eggs laid on each A4-sized sheet of paper, counting around 25,000 silkworms. Each silkworm needs to consume 20–25 grams of mulberry leaves in its lifetime, and for a sheet of silkworms, farmers have to provide 500–600 kg of leaves in total. The demand for food varies significantly at different stages of a silkworm's life. An ant-sized first-instar larva asks for very few chopped leaves per day, while at its fifth instar, it needs ample leaves for six to seven days in a row. Hence the intensity of a farmer's leaf-picking labour differs with the growing pace of the silkworm. Farmers accordingly decide how many sheets of silkworms to keep in each batch, measuring the size of their house and the number of leaf-pickers in a family. In Village Shangping of Yizhou County where I lived for months doing fieldwork, villagers raised one to three sheets of silkworms per family, with two sheets being the most appropriate and three very burdensome. At the busiest time of one batch, a family that hosts one sheet of silkworms needs to pick 100 kg of leaves per day and pack them into the rectangular plastic woven bags used to hold chemical fertilisers (Figure 2.2). Depending on

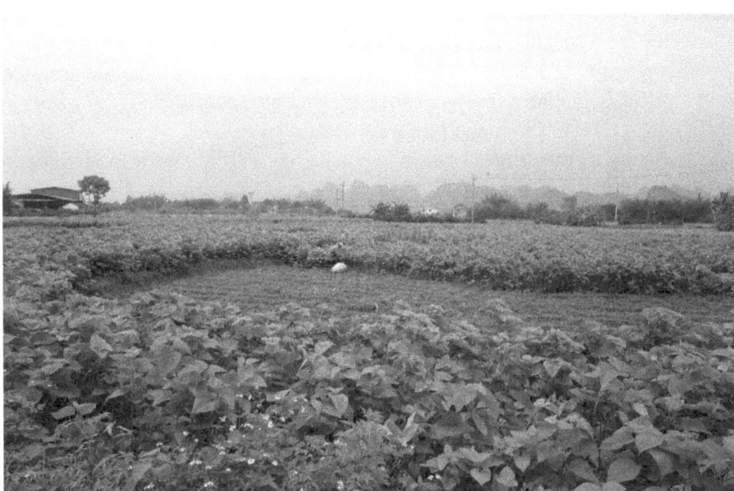

Figure 2.2 Mulberry fields at Village Shangping in Yizhou County, Hechi, Guangxi. Photo by author

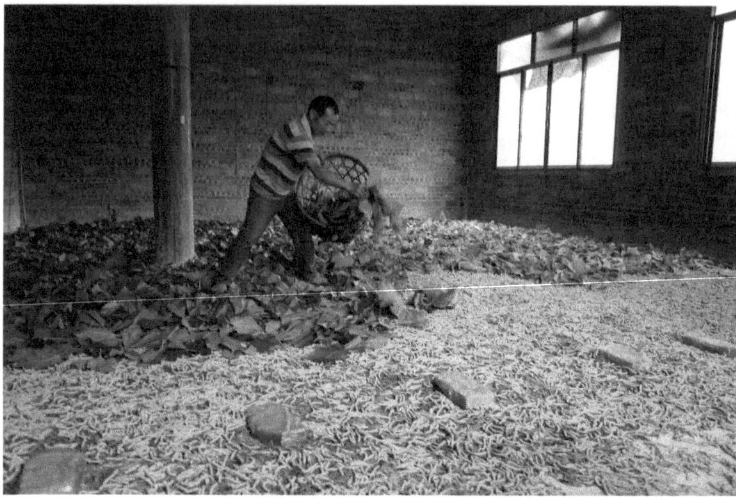

Figure 2.3 Farmer feeding silkworms in their farmhouse. Photo by author

how lightly the bags are filled, one bag of leaves is 10–20 kg and normally it takes one person more than one hour working in the field to fill one bag.

There are two 'workplaces' for silkworm farmers, namely, the mulberry garden in the field and the bedroom for silkworms in their farmhouse. The majority of local villagers in Guangxi keep silkworms within individual homesteads not far away from their mulberry tree lands, and the spaces spared for silkworms are typically the floor of the living room or an empty bedroom neighbouring the bedroom of family members (Figure 2.3). The labour in the mulberry fields includes fertilising, disease and pest control and, most importantly, leaf picking. Once the mulberry leaves have been transported to the house, usually by a family-owned mini truck, they are cut into strips and fed to younger larvae, while silkworms of the fourth and fifth instar are large enough to chew the whole leaves. Keeping the indoor environment clean is generally considered a household chore, but it is also part of the sericultural labour. As silkworms are extremely susceptible to various pathogens including bacteria, viruses, and parasites, a hygienic and well-ventilated room is essential to keep them healthy. However, among all these tasks in a sericultural worker's routine, leaf picking is still the most physically demanding one.

Silkworm farmers in China today, like their predecessors in pre-industrial ages, still practise hands-on husbandry that has not yet been largely mechanised into assembly lines unlike some other livestock husbandry. The main obstacle lies in the mechanisation of caring for the mulberry trees (Zhao 2005). In the development of sericulture in China, there are not yet enough human and material resources invested in the invention of leaf-harvesting machinery and equipment. Existing leaf-peaking machinery cannot distinguish between young and mature leaves in operation; nor can they avoid muddy or yellowed leaves which would contain fatal pathogens for silkworms. Automatic shears can also cut off branches and thus affect the number of leaves drawn on the trees afterwards. Therefore, in most rural areas in China, mulberry leaves for sericulture are still largely picked by hand.

It should be noted that another route to sericulture automation exists by replacing fresh mulberry leaves with artificial fodder. Research on silkworm fodder has been a significant trend in sericulture science study, triggered by high-end market demand where silk can be the raw material for the production of cosmetics, medical materials (stitches, cytoskeletons, and intravascular stents) as well as pharmaceutical proteins. These products have extremely strict quality control standards for ensuring human health and wellbeing. As various microbes and heavy metal elements may be left in mulberry leaves, leaf-fed silk can hardly reach these safety requirements. On the other hand, fodder is much more controllable than mulberry leaves with regard to 'purity' and 'sterility'. Japan therefore witnessed the world's first success in artificial fodder-fed silkworm raising during the whole larval phase in the lab in 1960 (Fukuda et al. 1960). But, as the industry of raising silkworms for silk fabrics productions in Japan has already shrunk, in recent years, small-scale silkworm farm factories have sprung up in the country to produce high-end silk materials, where silkworms live on fodder for their entire larval stage from the first to the fifth instar before cocooning. Fodder-fed silk products can fetch high prices in the market of personal care products and medical supplies, making up for the expensive cost of fodder costs.

It was during my second trip to Guangxi Sericulture Research Centre in Nanning City on 4th April 2019 that I interviewed the chief biologist, Mr Zhang, at this institute, who is the leader of a research team on silkworm fodder study. On that day, I visited their lab when some biologists there were conducting a comparative experiment on three kinds of fodder made in China and Japan, respectively. The tiny second-instar silkworm larvae were kept in lunch-box-sized plastic containers with the man-made fodder laid in the bottom, like a clump of brown moist clay (Figure 2.4). According to Zhang, fodder-fed silkworms are smaller and produce much finer but thinner silk than those leaves-fed. Thinner silk filament with better purity is less suitable for reeling and weaving into fabrics because it breaks easily on existing reeling machines in silk factories. Moreover, although silkworm fodder is expected to provide essential support for potential large-scale labour-saving silkworm husbandry without seasonal restrictions, the cost is too high for farmers to make money compared to the current hands-on mode. As he explained:

> If rearing silkworms with fodder in the whole larval phase like Japan, using three and a half catties of fodder can harvest about one catty of the cocoon. The purchase price of cocoons in Guangxi lingers around 20 yuan

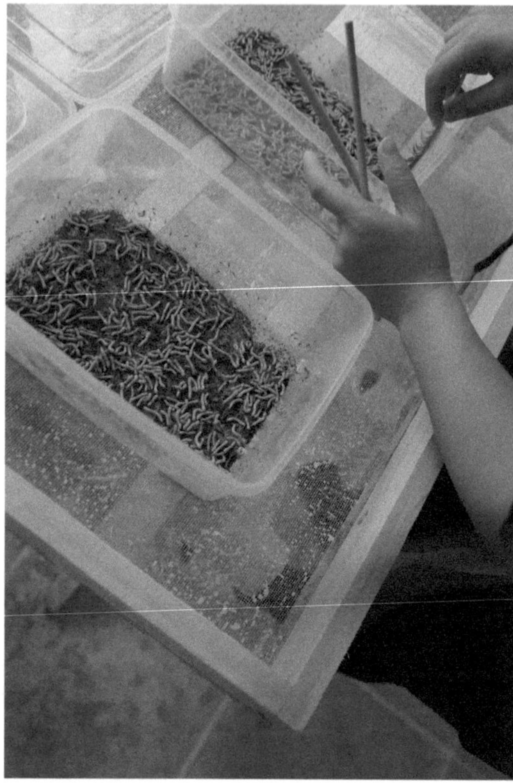

Figure 2.4 Second-instar silkworm larvae on artificial fodder in Zhang's lab. Photo by author

> per pound, while the cost of one catty of silkworm fodder is also 20 yuan. It would not be economical to raise fodder-fed silkworms for traditional silk fabric productions. Should there be a breakthrough in mulberry care mechanisation, it would be helpful to scale up the silkworm husbandry. But it is impossible to make money when you go for scale with the present model such as all-age fodder-fed farming.

The biologist further compared the technological 'backwardness' of sericulture from the perspective of industrialisation to the heyday in ancient China. Throughout the civilisation of China, silk had been a significant vehicle of material culture. It had economic and material importance in the imperial era, facilitating cultural communication and being used to pay taxes. Even during the early years of the establishment of the People's Republic, the country earned foreign exchange by exporting silk, which was then used to buy high-tech weapons and advanced equipment (Zhao 2005). However, the fact that silkworm cultivation remains labour-intensive demonstrates that 'its progressing pace cannot keep up with that of our society as a whole', as Zhang noted, and as silk has lost its previous importance, sericulture cannot return to the past heights.

Sericulture in China can, nevertheless, be regarded as modern as it is deeply enmeshed in the logic of the capitalist market. Although farmers' sharing space in the farmhouse with silkworms indicates affective and physical proximity, their cohabitation and coordination of biological clocks are not immune to the forces of capitalism. The aim of this human-silkworm symbiosis for silkworm farmers is to earn 'quick money' from selling cocoons, an expression I heard many times in interviews and chatting in Village Shangping. The relatively short circle of investment and return of silkworm husbandry corresponds to the whole life cycle of domestic silkworms lasting merely six to eight weeks. The economic risk of each batch in raising silkworms is thus manageable. On the other hand, local cocoon purchase prices fluctuate with the global raw silk market. The standard of regulating time in this framework is uniform. Synchronised pacing enables easy management, saving in labour and resource input, and hence better accumulation of profit and in the case of sericulture specifically, accumulation of bio-capitals. Farmers, therefore, apply hormones to regulate the growth of silkworm larvae for better efficiency. Although farmers and silkworms are not regulated in a conventional apparatus of unified time management such as those in a factory farm, bodies of humans and nonhumans are caught in the rhythms of the global capital market. However, silkworm cultivation might count as 'premodern' from the perspective of technological progressivism. The silkworm as a domesticated cash animal has colluded with the mulberry tree to resist the tide of alienation in automated farm factories (in Marxist terms) in a capitalist framework. This silkworm-mulberry coalition instead asks for the hands-on care of their human feeders.

Silkworm husbandry hence traverses ancient agrarian practices and modern industrial agriculture. This liminal quality of the sericultural industry has been largely ignored in understanding the critique of the accelerating industrialisation of contemporary society in Liang Shaoji's silkworm art projects. In Liang's installations such as 'Chains: The Unbearable Lightness of Being/Nature Series No.79' (2003) (Figure 2.5) and 'Snow Cover' (2013), iron chains and plastic products signify the burden and redundancy modernity has brought on humanity, while the silk flakes covering those objects carry the power of '*ziran*', denoting a natural, organic, and primaeval and harmonious flow (e.g. von Drathen 2011; Xia 2015). The silk-silkworm assemblage works as a symbolic metaphor for '*ziran*' as nature because these critics hold the underlying assumptions of a dichotomy between both natural animals (silkworms) vs. inanimate objects as well as organic material

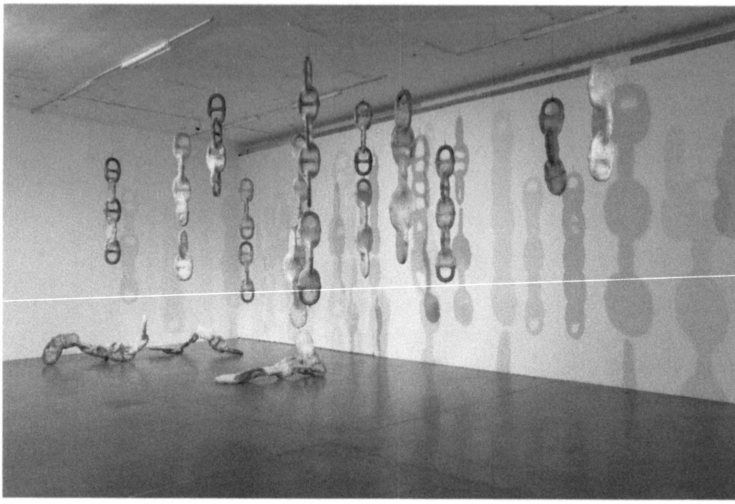

Figure 2.5 'Chains: The Unbearable Lightness of Being/Nature Series No.79', 2003. Copyright: Liang Shaoji and ShanghART Gallery

(silk) vs. machine-made products. Nonetheless, as demonstrated above, silk and domestic silkworms contextualised in the current sericultural industry in China should not be considered the antithesis of either industrialisation or modernity, but rather this organism-material combination presents a state of transition, or liminality. It is the non-binary relationship between domestic silkworms and their human cultivators that facilitates Liang's convoluted criticism of contemporary society in China. Silkworms are domesticated but silkworm husbandry is not yet automated. Silkworms are domesticated but silkworm husbandry is not yet automated. The circulation of silk has been internationally marketised, but the non-human silk producers are still raised within domestic intimate spaces in most cases in China. This subtle duality in silkworm husbandry functions as the ground of Liang's artworks where his juxtaposition of silk and other man-made or manufactured objects can afford complex symbolic articulation and functional purposes (more discussion in Chapters 3 and 4).

Liang as the Cultivator

From 1989 to 2000, Liang spent nearly a decade living in a small village near Linhai County in Zhejiang Province near the coast of Eastern China. While feeding silkworms and growing mulberry trees with local farmers, he observed and learned about the physiological and behavioural characteristics of silkworms through his daily labour. Liang also photocopied textbooks from the sericulture department in then Zhejiang Agricultural University and subscribed to academic journals such as 'Science of Sericulture' to follow up on the latest research findings as another approach to equip himself with sufficient knowledge. In these early years, Liang took care of his mulberry garden and silkworms by himself and even attempted to build greenhouses for bush-type mulberry trees, hoping to get around seasonal climate restrictions on the length of silkworm farming season. However, these were often destroyed due to the high frequency of typhoons in this region during conventional silkworm seasons in late spring to summer. After ten years' of intense engagement in silkworm husbandry, Liang's focus shifted more towards art creation and his limited time and energy made it difficult for him to continue juggling the heavy workloads of leaf-picking and worm-feeding as before. Liang then started to purchase mulberry leaves from local farmer acquaintances and hired them to look after the silkworms before the larvae entered the fifth instar to be transported to the artist's studio.

On a late night in early April 2019, I spoke with Liang Shaoji in his Tiantai County studio, where he shared technical insights from his decades of agrarian work. The silkworm larvae fed heavily on mulberry leaves during their fifth instar, but Liang noticed that if the leaves were very wet the silkworms would urinate more before cocooning. Feeding drier mulberry leaves then can slightly reduce Liang's workload in creating his art installations as the likelihood of the silk flakes being contaminated with an extra spill of urine would be minimised. Mature silkworm larvae in their late fifth instar in farmhouses climb up to the cocoon frame to prepare for cocooning. As they crawl along, the insects would empty their bodies of urine and faeces while secreting silk filaments before substantially building ellipsoidal cocoons. Even if a very small number of silkworms are excreting within the cocoons, those contaminated ones usually do not affect the sold price of the whole batch of cocoons and therefore farmers are not too concerned about silkworms' excretion. However, as Liang planned for the silkworms to spin 'flat cocoons' on planes of various materials, the lightest urine stain would detract from the flawless whiteness of the entire silk flake. Hence the artist would rather waste the initial silk that silkworms had just started to produce. He took great pains to observe each silkworm's body in the light to make sure those larvae had excreted up before placing them on a pre-arranged assembly of objects to be wrapped in silk. 'It all depends on feeling on fingertips and experience', Liang said in response to my question about how to tell when silkworms' excretion is complete.

The difference in perspective between the artist and the farmer is also reflected in the attentive observation of the movement of silkworms. In Liang's studio, I saw several mirrors on the workbench covered with thin sheets of flat silk cocoons. The bright, clean, and shiny surface of the mirror allowed me to easily identify the figure-of-eight trajectory of the silkworm swaying its head as it spun out the single silk filament. I also perceived that the density and texture of these silk flakes on each mirror varied quite significantly. The artist explained why the silk traces take different forms:

> The silkworm movement changed over time during its cocooning phase. In the first two days when they just started secreting silk, they were more excited and active, so their heads bobbed more wildly. Later in the cocooning process, they got a bit tired and thus their heads oscillated in a narrower range. The texture of each silk flake on a flat surface will manifest this difference in movement accordingly.

Liang also noted an anecdotal episode that happened during his creative process using the 'flat cocoon' technique:

> Mr Zheng, the chief biologist at local silkworm breeding station in neighbouring Linhai, provided me with a lot of support with knowledge and equipment during my artistic experiments with the silkworm, and once he came to my studio for a visit and advised me to pad cotton on the flat plane to absorb the silkworm's urine. We tried, and the cotton and silk flakes resulted in tightly gluing together, even harder to dry and clean.

Farmers would not be able to see with the naked eye the movement and metamorphosis of silkworms as the larvae gradually wrap themselves in thicker and less transparent ellipsoidal cocoons in contrast to Liang's flat planes. Liang's acquaintance with the behavioural traits of spinning silkworm larvae on 'flat cocoons' (see more technical details in Chapter 3) therefore could never have been empirically gained by the farmers engaged in traditional sericultural production.

As Liang mentioned, the government-sponsored Silkworm Breeding Station in Linhai provided technical and equipment support for the artist's breeding experiments to fulfil his artistic ambition. At the institute, silkworm scientists conduct breed selection most often to improve the efficiency and productivity of silkworm husbandry. After the theoretical framework of modern sericulture and breeding techniques had been introduced to China from Japan in the early twentieth century, local silkworm varieties kept by farmers were collected by sericultural research institutes across the whole country. After national validation, these 'breed materials' constitute the national breed bank of silkworm varieties which are then selected and improved by scientists to suit diverse production needs of the industry. As a result, there are innate physiological differences between lines or varieties of the silkworm in their responses to various pathogens, known as tolerance or even resistance to a certain disease. A variety with good tolerance to one pathogen means a smaller percentage of infections, less sensitive reactions, or less severe symptoms.

I was told by Mr Zhang during my stay in the Guangxi Sericulture Research Centre that the selection process for disease-tolerant silkworm varieties is measured in years. Breed selection takes too long for the lifespan of a disease-ridden silkworm, and it is also too limited and feeble in regard to the history of a species in concert with other organisms. The purpose of breeding, in Zhang's words, is to 'solve problems and needs in human's silk production'. This antidote, however, is often lagging behind epidemics; it mainly serves a future time, in the name of long-term development and benefit of the industry. If breeding for tolerance to a long-standing enemy like pébrine disease is less urgent and thus seems tardy, the selection work by the biologists for even newer threats in production can also be time-lagged, which is well manifested in the case of the selection of fluoride-tolerant varieties.

Fluoride, a non-biotic pathogen of silkworm disease, became a threat to the sericultural industry when the waste gases from the factories containing fluoride dust which fell on the mulberry leaves in the fields nearby. When a silkworm consumes contaminated leaves, the concentration of fluoride in its body interrupts its metabolism, causing black scarring on its skin and then death before pupation and cocooning. Mr Zhang briefly recollected the history of breeding fluoride-tolerant varieties. Japanese scientists then considered breeding varieties with stronger fluoride tolerance to be raised in areas with highly developed industries where local farmers still needed to rear silkworms. They did find some varieties desired eventually, yet applicable regions with the need to develop both sericulture and industry simultaneously became rare. There are more direct and quicker ways to solve the air pollutant poisoning than through breed selection for tolerance, such as using administrative means to reduce the fluoride emissions or even relocating the factories or mulberry farmlands. Often, where the industry is well developed, agriculture is inhibited; factories and mulberry fields do not always exist next to each other.

Nevertheless, fluoride poisoning was indeed a serious challenge in sericulture a decade or so ago in both Japan and China. Liang Shaoji mentioned a local fluoride pollution accident he was involved in.

> I was living in a village in rural Linhai. Next to the village's farmlands, there was commercial land. New houses were being built over there and farmers went to pick mulberry leaves in the fields beside the construction site. The silkworms all subsequently died. Later I helped the local farmers file a lawsuit for compensation...Our human history is too similar to that of the silkworm. The domestic silkworm has a history of more than three thousand years, about the same as that of Chinese history.

Liang's last remark points to the similarities both in the length and undergoing transformations in two species' history on the same planet. It suggests that ecological challenges (if not disasters) brought about by modernisation projects have affected not only human communities but have forced animals to adapt accordingly, just as the fluoride case shows. As a pollutant fluoride damages the bodies of humans and silkworms alike, individuals either get sick and die or adapt with better tolerance to survive.

Liang's original aim of breeding selection was to find varieties strong and tolerant enough to be able to spin a 'flat cocoon' on metal which in the artist's view best represents cold-hearted industrialisation and modernity in both material and symbolic terms, as he stated. To counter the resistance of silkworms to unfamiliar materials such as metal and human skin, the biologist Mr Zheng and his colleagues assisted Liang to hybridise silkworm varieties among breeds in Zhejiang and from other provinces. After a number of trials, Liang eventually selected a few promising 'candidate' breeds with stronger physical strength and higher immunity to continue his artistic experiment. Liang offered more details on his hybridisation experiments:

> I have compared both hybrid and purebred varieties when making artworks and prefer the hybrid ones when dealing with metal materials. Purebred silkworms are more delicate, producing good quality silk but not sturdy enough to climb up high. It is probably like interracial marriage with different genes mixed. A multiracial baby would then have distinctive physical traits.

Whether accurate or not in terms of genetic science, Liang's anthropomorphic analogy between hybrid silkworms and people with mixed races tends to emphasise a kind of cross-species continuity between humans and silkworms. For Liang Shaoji, this analogical continuity manifests itself in the adaptive strategies of life in ecological history at the diachronic scale, as in the case of fluoride poisoning, and also in the biological principles at the synchronic scale (see also Chapter 5).

Turning back to Liang's hybridisation trials, those varieties with no significant advantages for cocoon production selected from the artist's continuous adjustments in the experiment enabled him to initiate generative encounters between silkworms and different material surfaces and subsequently succeeded in realising his artistic vision. Liang's activities in the laboratory constitute part of silkworm cultivation, i.e. mating and crossbreeding, and at the same time part of his artistic creation. Throughout his career, Liang's artistic techniques with silkworms have been characterised by this intricate interplay of 'making-in-cultivating', or 'cultivating-in-making' – a nuanced approach exemplified in projects like 'Listening to the Silkworm', which I will explore in more depth.

Liang as the Maker

The tricksy remark – *'We delight in it, so do silkworms'* – uttered by a Linhai farmer about silkworms' potential enjoyment of local opera music intellectually chimes in with anthropologist David Graeber's (2014) reading of the

Daoist story 'Fish's Pleasure' in '*Zhuangzi*'. As recorded in this ancient text, Zhuangzi and Huizi, two Chinese philosophical personae, were debating on the dam whether it would be possible for them to claim that the fish drifting in the river were happy. When questioned by Huizi 'you are not inherently a fish then how can you know a fish's pleasure', Zhuangzi acknowledged his vantage point as 'knowing from my standing on the bridge' instead of diving in the water. Zhuangzi's distanced but reflective positionality, as noted by Cantor, can be understood as a 'species-specific perspective', suggesting 'the limitedness of human knowledge' (2020: 221). Graeber, however, tactically avoids the discussion of execution of empathy across different species but instead turns to the sameness, the ubiquitous activity of having fun 'for the sheer pleasure of doing it': 'We can understand the happiness of fishes – or ants, or inchworms – because what drives us to think and argue about such matters is, ultimately, exactly the same thing' (2014). Silkworms in Linhai, in the eyes of local farmers, are also capable of having fun, and more importantly, capable of having auditory joy in Yue opera. Different from Zhuangzi and Huizi's speculation over other creatures from a certain standpoint, the farmer Liu's expression suggests an affective interspecies sociality: I gain pleasure in Yue opera. You silkworms stay with me in the same space, surrounded by this melody too. I do sincerely hope and believe that you also feel happy as I do. Here the spectre of the ancient Chinese sage Mencius is haunting over the village near Linhai, whose exhortation to the emperor later became a popular Chinese proverb: 'Pleasure (in music) alone is not as good as pleasure for all (独乐乐不如众乐乐)'. Sharing is always an effective and engaging practice to firmly bond a relationship even between silkworms and farmers.

The artist was also inspired by that comment and conducted experiments on silkworms' response to music. He intentionally curated the repertoire for his artistic experiment with silkworms. Light music, usually without lyrics, also known as mood or ambient music, is regarded as easy to appreciate and requires less sophistication. It seems to be less culture-embedded than regionally popular genres such as Yue opera favoured by many villagers in Zhejiang. Liang chose this more accessible form of music (by human standards) in his trials to appeal to a possible wider silkworm audience:

> It was nearly twenty years ago when I only had a semiconductor radio, no smartphone. Some 'Light Music' programmes were regularly played on the radio. When I fed the silkworms, I placed the radio on the ground beside a rack of bamboo shelves, with pieces of light music live on the broadcast… However, the time span of my attempts was far too short to show any significant result on whether some light music would be of benefit to the silkworm's growth and wellbeing. Even biologists need to carry out a rigorous set of procedures in long-term controlled experiments. A scientifically solid outcome should be backed up by reliable data. It's not that simple for me.

Domestic silkworm is very fragile and sensitive to its living environment and susceptible to factors such as room temperature, humidity, air quality, the nutrition of the leaves and various pathogens invisible to the naked eye. It is always impossible for farmers to identify a certain culprit for a batch of poor cocoon yield or their insects' bad health. The same obstacle also lies in the artist Liang and scientists' endeavour to design and conduct experiments on silkworm's musical taste. After a field trip in Zhejiang Province, I conducted an interview with biologist Mr Zhang at Guangxi Sericulture Research Centre and asked about silkworm's auditory organs. I was told that silkworms don't have 'ears' to decode the 'sound wave', but they can 'feel' and react to the vibration in the air by receptors on their skin.

> It's more like our human beings' sense of touch rather than perceiving by an auditory sensation, and we still have very limited knowledge on the range of frequency and intensity of sound waves that are detectable by the silkworm. However, we found that when silkworms were eating the leaves and a person knocked on the shelf they were on, the larvae would become so nervous that they would stop eating.

Zhang showed unexpected interest in this topic of silkworm hearing and continued to recount a failed experiment that has never been recorded in any Chinese sericultural history book or academic journal:

> In the 1970s and 1980s, some scientists at the Sericultural Research Institute, Chinese Academy of Agricultural Sciences (SRICAAS) did an experiment on whether silkworms' listening to Peking opera can increase the cocoon yield. The result of that project, however, after long and careful deliberation, is still 'hard to tell' (不好说).

The early 1980s is a remarkable period in the history of science in China. The Cultural Revolution began in 1966 and lasted for ten years till 1976, causing a devastating blow to many intellectuals and scholars' academic research as well as their personal lives. In late March 1978, the National Science Congress was held in Beijing where Deng Xiaoping delivered a speech to advocate that 'the key to modernization is the modernization of science and technology'. The event marked the commencement of the 'Spring of Science', an ideological slogan derived from the closing speech title of that congress given by Guo Moruo, the president of the Chinese Academy of Sciences. The recovery of Chinese academia, especially the science community, had gained strong national support in the tide of modernisation and globalisation boosted by Deng Xiaoping's 'Reform and Opening-Up' policy since the late 1970s. The Sericultural Research Institute in the Chinese Academy of Agricultural Sciences (SRICAAS), founded in 1951, is a national research institute on sericulture located in Jiangsu Province. It seemed not a coincidence that Peking opera was the deliberately chosen genre in an experiment on possible correlations between the music and silkworm's wellbeing designed by Chinese biologists at SRICAAS in the 1970s and 1980s. Peking opera used to be favoured by the Qing court and later gained its state-wide long-lasting reputation as treasured cultural heritage of China. During the Cultural Revolution, Peking opera had been twisted in forms and themes into 'Revolutionary Operas' by Jiang Qing for the Communist Party's propaganda agenda. The exact timeline of these inquiries into cocoon harvests and Peking opera remains unclear, making it uncertain whether the tested silkworms heard the authentic or adapted version. Yet, the experiment's design still reflects the ideological underpinnings of Peking opera in its historical context.

The correlation between silkworm's health and certain types of music is impossible to be scientifically verified, as there would be far too many interfering elements to do a controlled experiment. Thus, it is hard to draw any solid result. Although an investigation into silkworms' musical preference tends not to be a promising research project within the bio-scientific framework, the motives behind those inquiries mentioned above remain noteworthy. The choices of music genre for silkworms by scientists, the artist, and farmers range from culturally loaded Yue and Peking opera to less-serious Light Music. As for the research project at the national-level Sericultural Research Institute, selecting Peking opera for silkworms to listen to, as previously noted, is a mild way to follow the mainstream national ideology at the early stage of the revival of Chinese technology and science study in the 1980s. In Linhai farmers' mindset, silkworms might be enjoying the Yue opera during their mealtimes when their human feeders turned on the radio or tape recorder and sprinkled the chopped mulberry leaves. This piece of local knowledge about silkworms emerged through their everyday practice of intense care and interaction which refuses any trans-local articulation. For Liang, doing an experiment on light music and silkworms' reaction is to go beyond the orthodox genealogy of sericulture study with his artistic imagination and sensitivity; at the same time, according to his own statement, it might also 'open up silkworm's capacity or nature (蚕性)' to appreciate some music composed by human beings. For an artistic trial, a lack of significant difference does not mean a failure of an experiment, as a 'scientific failure' can serve as another 'result' to ponder on for further artistic creation.

The unresolved artistic experiment with music and silkworms did not stop Liang Shaoji's meditation on auditory sensation and this insect, but years later, he took a path of the opposite direction to address this issue in 2006. If silkworm's response to music is inaccessible to human knowledge, humans can stop for a while and listen to the 'music' made by silkworms. 'Listening to the Silkworm/Nature Series No. 96' is a sound art installation that invites the audience in the gallery to wear headphones and listen to the voice recordings of silkworms' chewing mulberry leaves and also spinning silk filaments. This art project was first launched in RNC Art Museum

(Nanjing, 2006) and later in Zendai Museum of Modern Art in Shanghai (2009), as well as the True Colour Museum (Suzhou, 2010) and Hayward Gallery (London, 2012). In the last two venues, live silkworm larvae were transported into the gallery space and the sound of silkworms' chewing mulberry was also transmitted directly into the audience's headphones with the pre-recorded audio tracks.

At the 2nd Wuzhen Contemporary Art Exhibition in March 2019, Liang delivered a speech on his whole artistic voyage with silkworms. Liang described how he recorded the sound of silkworms at his studio in Linhai and explained the nuanced variations in the sound recording:

> Late at night, the room was filled with the sound of hundreds of thousands of silkworms spinning silk. It sounded like the falling raindrops, the babbling brook, in a mood of ancient Chinese landscape ink paintings. It is breath-taking. I tried to record this collective sound around midnight as it was the most silent period of time in the village. People were asleep, but dogs occasionally barked which jumped into the pure sound of silkworms. I had to restart my recording work again and again… The sound of silkworms' spinning can be divided into two types: the moment when the silk is secreted out, and the sound when the silkworm glues two silk threads together, resembling the sound of plucking strings.

Surprisingly, I noticed the artist's backstage execution of this sound artwork was almost identical to a piece of my own fieldwork experience. I, too, when first standing in a room full of fifth-instar silkworm larvae biting the leaves, was struck by the clear, constant rain-like poetic chords. I took out my voice recorder from my pocket, a convenient device I always carried to record my interviews with villagers in Yizhou County. It was in the daytime, so the soundscape outside that small room for silkworms was much more turbulent and vigorous than that at midnight. Bird's call, dog's barking, the roaring of a motorcycle engine, and a child's crying randomly joined the chorus, but I was not annoyed by the hubbub of these uninvited matters. Unlike Liang, who might need a purer and more ethereal audio track to involve the audience in the meditation on the silkworm alone, I recorded the sound in situ just to recall this moving moment for myself and maybe later, to share it with more people who would not be unable to listen to this tranquil 'music' live. Hence, I don't mind the messiness, the entanglement of different pitches, and the lively energies surrounding the silkworms and myself that were being documented in my voice recorder.

It is interesting to note when sericultural science didn't provide a satisfactory answer to Liang's artistic inquiry on silkworms' auditory sense, he turned to ancient Chinese philosophy to seek cultural nourishment on this topic. Liang confessed to me that he felt close to Daoist thought and pursued the spirituality of '*Tian-ren-he-yi*' (天人合一), 'the integrity or oneness of humanity and heaven/nature' in his art creation. In his artistic statement for 'Listening to the Silkworm', Liang (2014) emphasises the Daoist maxim 'the ultimate principle is to follow the way it naturally goes (*dao* follows *ziran*)' that is underlying this project and claims that he had practised this principle in his sound recordings of silkworms' eating leaves and spinning silk. Although Liang would wait for the sudden barking of the village dogs to cease, again and again, before pressing the record button, he did not manipulate or process the original sounds made by the silkworms (although he specifically prefers hybrid breeds). The artist did not alter the soundtrack or change the environment in and outside his studio. He just waited with patience till the dogs and frogs in the village closed their mouths, and then he could turn the microphone to silkworms, the designated soloists for this art project.

While the ancient Daoist philosopher Zhuangzi proposed in the fourth century BC that you 'listen with your heart, not merely with your eyes' (无听之以耳而听之以心), a contemporary anthropologist Ingold (2000: 285) articulates a parallel concern that humans should not only make sense of the senses as cultural constructions of sensory experience, but should dwell in the environment, sense the world, and make sense of ourselves to others. Silkworm and humans do have 'auditory' experience though with divergent sensory apparatus and thus as cohabiting partners can listen to ambient sounds together such as the Yue opera or light music. Meanwhile, they might listen to and correspond to each other, as in the setting of '*Listening to the Silkworm*'. The way Liang was making audio tracks in the village he lived in for this project was natural and spontaneous, as silkworms being recorded did not have to make extra effort to produce 'silence' or any uncanny sound but just eat leaves and spin cocoons. Despite that, the process of preparing this soundtrack can be regarded as artificial, unnatural, and manipulating. Unlike the rowdy polyphonic voice recording made during my fieldwork in rural site, the artist deliberately stripped away the noises made by other creatures as the ambient sounds of the village in Linhai. The finished recording of silkworms' solo is thus purified out of its context to create a more abstract, airy, and spiritual aura. Again, the artistic strategies employed by Liang here blurred the technical boundaries between cultivation and making, between '*ziran*' as spontaneity and determined interventions.

Techniques of Making and Cultivating

In art criticism, both aesthetic and semiotic readings of Liang Shaoji's artworks (e.g. von Drathen 2011; Gladston 2011; Xia 2015) focus too much on the finished art projects at the expense of the processes, such as the intense silkworm cultivation labour and biological crossbreeding experiments delineated in this chapter. It is also largely neglected in interpreting Liang's art that silkworm husbandry in China sits at the liminal point among highly mechanised modern agriculture, the global silk market, the family farming model, and hands-on personal cultivation. This leads to the erroneous assumption that domestic silkworms and the silk they spin represent a primaeval and intact state of '*ziran*' uninfluenced by accelerated progressivism and thus in Liang's art projects form as the direct antithesis of industrialisation, modernity, or alienation in the capitalist system. Such overwhelming concentration on the final objects echoes Ingold's criticism of anthropological studies of material culture which spotlight what happens as artefacts 'become caught up in the life histories and social interactions of the people who use, consume or treasure them' but neglect the ongoing processes that bring the artefacts themselves into being (2013: 7). As for the silk-silkworm duo in Liang's works, the growth of silk and silkworms and the artistic techniques to interact with this species are effortlessly concealed by the end presentation of art installations and performances. The attention of both art critics and the audience is too often attracted by the finishing moment, the objective, or the fulfilment.

Although in the 'posthuman' literature, terms such as nonhuman and hybrid have become fashionable, as Hallam and Ingold (2014: 16) notice, in academic discourses on materiality, the artefactual is privileged over the organic, while the techniques of making over the techniques of growing. They thus offer a more insightful framework to understand myriad and fluid relations between organisms and artefacts to dialectically analyse the processes of 'making' and 'growing'. Making has been compared to 'a rite of passage' (cf. van Gennep 1909; Turner 1969), when a thing undergoes ontological transformation by the maker's agency and crosses 'a threshold' from preparation to employment (2014: 2). The material has been removed from its former settings to the studio or workshop to be processed, and then it moves and reincorporates into a new phase of life. Growing, on the other hand, is material accumulation to incorporate substances into a being's body as a process of 'self-making or autopoiesis' where the grower contributes to 'setting up the conditions' to make sure the growth can proceed (2014: 3). From this perspective, the techniques of cultivating silkworms and the process of making art in Liang's practice, as portrayed above, is never radically split apart.

In raising and feeding silkworm larvae, Liang gained great physical support from local farmers in Linhai village. While crossbreeding stronger varieties, biologists also assisted Liang in the lab operation. Silkworms hence were co-grown by farmers, scientists, and the artist together before they were moved to the artist's studio in their fifth instar. Within the art studio, 'a rite of passage' of both the silk and the silkworm took place. Silkworms left the agricultural context where they were supposed to produce cocoons for sale but instead entered artistic social relations as the prototype of silk threads. Silk threads, too, also experienced a morphological transformation to participate in constituting art installations in the shape of the flat cocoons. It can be said that Liang as a grower had set the conditions for the silk to grow, accumulate

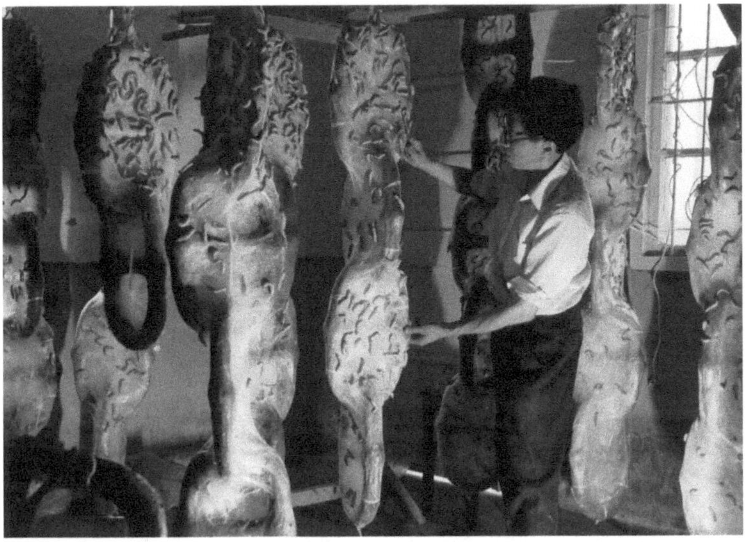

Figure 2.6 Liang Shaoji working on 'The Unbearable Lightness of Being' in his studio in 2002. Photo by Lihui Feng. Copyright: Liang Shaoji

and extend on other objects in his artworks while he also acted as a maker to foster a transition of silkworms' social role with humans (Figure 2.6). The confounded boundary between techniques of making (art) and cultivating (organism) in these silk-generating processes then challenges the ambiguous 'natural-ness' (*ziran*) in the essence of silk (as material) and silkworms (as life). It remains an issue whether Liang's art skills and strategies, including experiments with industrial-agricultural practices and crossbreeding, are consistent with the Daoist maxim 'to follow the way it naturally goes (*dao* follows *ziran*) [道法自然]' that he claims to adhere to. The next chapter will scrutinise more on liminalities in his art-generating processes by evaluating the specific technique of producing 'flat cocoons' to further complicate this question.

References

Cantor, L., 2020. Zhuangzi on 'Happy Fish' and the Limits of Human Knowledge. *British Journal for the History of Philosophy*, 28 (2), 216–230.

Field, J., 2014. Silk Production: Moths, Mulberry and Metamorphosis. *In*: E. Hallam and T. Ingold, eds. *Making and Growing: Anthropological Studies of Organisms and Artefacts*. London: Routledge, 25–43.

Fukuda, T., Suto, M., and Higuchi, Y., 1960. Silkworm Raising on Artificial Food. *Nature*, 187, 669–670.

Gladston, P., 2011. A Conversation with Liang Shaoji. *In*: *Contemporary Art in Shanghai: Conversations with Seven Chinese Artists*. Hong Kong: Blue Kingfisher.
Graeber, D., 2014. What's the Point If We Can't Have Fun? *The Baffler*, 24.
Hallam, E. and Ingold, T., eds., 2014. *Making and Growing: Anthropological Studies of Organisms and Artefacts*. London: Routledge.
Ingold, T., 2000. *The Perception of the Environment: Essays on Livelihood, Dwelling and Skill*. London: Routledge.
Ingold, T., 2013. *Making: Anthropology, Archaeology, Art and Architecture*. London: Routledge.
Liang, S., 2014. *'Back to Origin' Catalogue*. Shanghai: ShanghART.
Turner, V., 1969. *The Ritual Process*. Suffolk: Penguin Books.
van Gennep, A., 1909. *The Rites of Passage*. Chicago: Chicago University Press.
von Drathen, D., 2011. Liang Shaoji: Of the Immaterial and Its Corporeality. *In*: M. Bloemheuvel and T. van Kooten, eds. *Windflower: Perceptions of Nature*. Rotterdam: NAi Publishers.
Xia, K., 2015. The Silk Shadow of Chora-topia [online]. *ShanghART Gallery*. Available from: https://www.shanghartgallery.com/galleryarchive/texts/id/8092.
Zhao, F., 2005. *A Comprehensive History of Silk (中国丝绸通史)*. Suzhou: Suzhou University Press.

3 In the Web of Flat Cocoons

The visual-material pattern of the 'flat cocoon' dominates most of Liang Shaoji's art installations where thousands of silkworm larvae directly secrete silk filaments on pre-arranged materials covering them with cloud-like white semi-transparent silk flakes. Everyday discarded objects such as plastic cups, cigarette packets, and a keyboard are encased in silk in 'Snow Cover' (2013–2014). Meanwhile in 'Time and Permanence' (1992–2018) over a larger time span, batches of silkworms spun silk over pieces of pyramid-shaped barbed iron wires and previous layers of silk flakes over decades (Figure 3.1). Other installations have more specific referential implications. Miners' protective headgear and headlights in 'Helmets' (2004), for example, are intended to memorialise those who died in China's frequent mining accidents in recent years. 'Wenchuan Stones' (2014–2016) adopted a similar artistic 'syntax' where Liang transported the rubble from the ruins of the Wenchuan earthquake in 2008 to his studio to be enfolded by 'warm and soft' silk as a gesture to 'caress the wounds and comfort the afflicted', as Liang told me. Liang is concerned with not only social and environmental issues in China, but also international affairs as exemplified in the large-scale installation 'Destiny' (2012–2014). This work is set against the backdrop of marine pollution and the Gulf War (Xia 2015) composed of silk-bounded oil barrels and iron plates resembling the wreckage of sunken ships. Additionally, Liang has two performative artworks, namely, 'Self-roped' (2000) and 'Cloud Mirror' (2007) where the implementation of the flat cocoon technique also pervades.

The visual presentation of silk flakes as the wrapping material in these artworks, according to many art critics (e.g. von Drathen 2011; Xia 2017; Yang 2018; Zulueta 2016), represents the force of '*ziran*', a healing, comforting, protective and energising power from nature. Nonetheless, how such a metaphoric relationship between silk and '*ziran*' can be validly established in Liang's artistic expressions is still under-examined when these projects are categorised and interpreted as installations and the audience's attention is thus driven to the finished art objects (as discussed in Chapter 2) rather than the generating processes before their entering into exhibition spaces. This chapter will pave the way to address this gap by closely examining the details of the

DOI: 10.4324/9781003569091-3

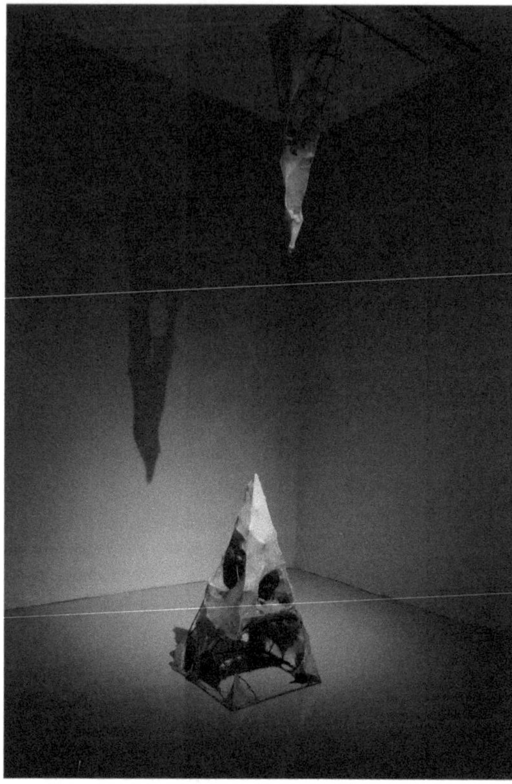

Figure 3.1 'Time and Permanence', 1992–2018. Photo by author

'flat cocoon' technique, whose artistic application contrasts with its employment in agricultural contexts.

The Flat Cocoon Technique in Sericulture

Silkworm cocoons are ellipsoidal in normal circumstance, though under human intervention, cocoons can also be flat. Cocoon-making is to protect the silkworm during its metamorphosis from the stage of larva to pupa, as the pupa loses its mobility and therefore is more vulnerable to natural predators such as bees, birds, and beetles, or pathogens carried by other insects in the wild. Hidden in an ellipsoidal cocoon shell, the pupa can more safely survive into a silkworm moth from the adversities of being eaten or getting ill. When the pupa becomes a mature moth, it will secrete proteolytic enzymes that can dissolve the glue wrapping the single silk thread as a cocoon and fly out. The empty cocoon left with a hole is thus destroyed into silk segments that are impossible to be reeled into long and fine silk filaments, resulting in their much-reduced economic value in silk fabric production. Domesticated from wild silkworms, they are often kept in farmers' farmhouses in Guangxi, China, where the threat of natural enemies becomes less lethal. Still, silkworms in thousands of years' sericultural husbandry retain their biological instinct to make ellipsoidal cocoons. Silkworm farmers have invented many tools to place mature fifth-instar larvae who are ready to secrete silk, varying from straw stacks to woven bamboo shelves and wooden lattice frames. The prepared silkworms will shake their heads in a figure-of-eight trajectory to choose a preferred narrow three-dimensional space to plan their cocooning strategies by locating the pivot points and pinch angles, and then spin silk threads from the outer layer to wrap themselves.

The cunning trick to induce silkworm larvae to instead spin 'flat cocoons' is to place them on a flat plane during the whole silk-secreting period where silkworms can find no pivot to make conventional ellipsoidal cocoons. Although it would be a strange or even worrying situation for a silkworm larva wandering on a level surface, the insect has no other choice but to produce silk regardless. One physiological principle of cocooning is the excretion of excess amino acids from the body of silkworm, which is the main ingredient for silk proteins, which means that silkworms otherwise would be poisoned to death if those various amino acids were not excreted within the given time. Silkworm husbandry already takes advantage of silkworm's excretion capacity, but the 'flat cocoon' is a much more extreme case of exploiting silkworms' biological traits where humans set up an 'unnatural' condition for silkworms to grow and produce. The technique was first applied in the silk industry in 1984 by Huang Canqi, the chief biologist of Chishang Sericulture Centre in Taiwan (Zhao 2005). Huang invented a wire mesh frame with very fine grids hanging at 30 degrees, where 500–600 larvae can climb and spin flat silk. By adjusting the angle of the frame's inclination, a uniform thickness of the silk pieces can be guaranteed while the inclined flat plane makes it easier for the spherical granular droppings of the silkworm to roll down to the ground. The silk flakes spun on flat planes are then be used to make silk quilts after simple degreasing and degumming processes, while conversely, traditional ellipsoidal cocoons need to be cooked, reeled, and then degummed before being processed into silk quilts, which is more time-consuming and labour intensive.

The actual implementation of this technique, however, was not as easy as it sounded. Climbing on inclined surfaces is dangerous for silkworm larvae as they would be more likely to fall off. Moreover, as silkworms will drain their bodies of faeces and urine before secreting silk to build cocoons, in normal sericultural scenes, the larvae's excrement will fall down through the space between hanging lattice frames without contaminating the ellipsoidal cocoons within each separated box (Figure 3.2). Nonetheless, when a large number of mature fifth-instar silkworms are placed on level surfaces regardless of

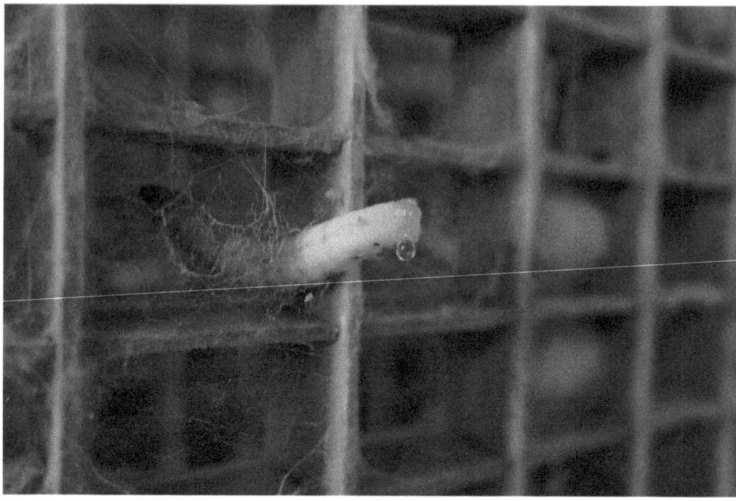

Figure 3.2 A spinning silkworm larva peeing outside the cocoon frame. Photo by author

whether their final excretion was completed or not, it is impractical for human workers to check the worms one by one. This uncertainty poses the risk of the white flat cocoons being contaminated with silkworm excrement. Though some farms in Taiwan later developed roller-shaped equipment to reduce the chance of silkworms falling and silk being spoiled, intense attention from human keepers is still required during the production of flat cocoons. Spinning larvae are phototropic and therefore move towards light. The rooms for making flat cocoons hence are always kept dark to ensure that silkworms are evenly distributed on the flat plane and thus the silk flakes are evenly thick. Farmers, however, have to spend long hours with the silkworms during the five or six-day silk-secreting phase, picking up the fallen worms and adjusting their density.

The advantage of using the flat cocoon technique in making silk quilts is that it can significantly reduce the industrial wastewater from boiling ellipsoidal cocoons and degumming silk threads. Extra chemical treatments as potential pollutants are also required to process round cocoons into silk quilts. Comparatively, turning flat cocoons into quilts by lye degumming is much simpler and less contaminating to the environment. Moreover, boiling ellipsoidal cocoons containing silkworm pupae inside tend to leave the smell and impurities from the worms on the silk, but the flat cocoons allow silkworms to moult and metamorphose into moths elsewhere after spinning silk threads and subsequent pupation. Silk quilts made from flat cocoons, therefore, are regarded as fluffy, soft, and aerated, with an additional economic advantage of leaving intact silkworm pupae for possible processing into food commodities.

Nonetheless, the sericultural industry in Taiwan has been decimated from its heyday in the 1970s, as Japan turned to mainland China to buy cheaper cocoons and raw silk since the early 1990s. The Taiwan government no longer provided state support to this industry and instead subsidised silkworm farmers turning towards other trades. Only Chishang Sericulture Farm and a few other farms in Miaoli County survived from the decline of sericulture in Taiwan, keeping the key technique of applying flat cocoons in silk quilt production. The 'flat cocoon' technique was also imported and adopted by some companies in mainland China to make silk quilts and surface materials for calligraphy and painting, but never replaced the mainstream ellipsoid-shaped cocoons in the silk industry.

The Flat Cocoon Technique in Art

Liang Shaoji has largely used this flat cocoon technique in his silkworm artworks. Despite his artistic visions being far removed from producing silk quilts, Liang shares the same worry about the excrement contamination on silk flakes when making most of his projects. Flat silk pieces should ideally be white, light, and thin to achieve a visually ethereal presentation. The artist, therefore, has to devote as much care, attention, and physical effort as silkworm farmers do to ensure the flat cocoon-making processes will be carried out in accordance with his expectations. To check whether each silkworm has completed its last excretion before making flat cocoons, for example, the artist needs to stay up all night by the fifth-instar silkworms as they are about to secrete silk so that the flat silk flakes on his semi-finished art installations can be pure and spotless (also in Chapter 2). A cocoon consists of only a single continuous silk filament. This fibre is made from a fibrous protein 'fibroin' in the core wrapped by glue proteins named 'sericins'. The sticky silk glues function in the formation of a cocoon by gumming the silk fibres together. Even the flat cocoons used in the production of silk quilts are subject to a degumming process, but Liang confirmed that silk flakes on all of his installations that have been openly exhibited have not been degummed:

> I did comparative experiments. Degummed silk threads can be extended and elongated, perhaps allowing for more variation in shape, but they are no longer as lustrous as before. The silk loses its gloss. Silk threads keeping their glue have a certain stiffness and are more likely to retain their original shape. Once degummed, they became too soft like a thin cloth. Hence I don't do much degumming of flat cocoons in my works.

For these reasons, to guarantee the flawlessness of the flat cocoons and their position and density in his artworks, Liang's daily schedule during the cocooning phase is very irregular. Staying up late during Liang's art processes has messed up his biological clock and damaged his health, the artist told me.

However, the implementation of the flat cocoon technique not only affected the health of the artist but also exposed the silkworms to a tougher environment to survive and secrete silk. While exploring the behavioural characteristics of cocooning larvae on flat planes, Liang also noticed a phenomenon that silkworms who secreted flat cocoons and went through pupation in the open air without the protection of ellipsoidal cocoons were less likely to complete their metamorphosis into adult moths and died abnormally. Liang worried about what he observed and hence went to Zhejiang Agricultural University to discuss with biologists in the sericulture department. Silkworm biologists who are less interested in the visual and material potentials of flat cocoons in art are instead concerned with the physiological conditions of the 'naked' silkworms on flat surfaces, namely, their sensitivity to the environment, susceptibility to pathogens, and the sharply reduced proportion of completed metamorphosis into moths. I asked Liang whether he obtained satisfactory answers from the experts. The professors were surprised, as Liang recalled, as the health of silkworms on flat cocoons was supposed to be a doctoral-level research topic in biological studies and Liang, as an artist, was unexpectedly knowledgeable about it. As both the artist and biologists have recognised, leaving silkworms 'naked' to spin flat cocoons against their instincts has posed challenges to their physical health. As Liang explained:

> I think it was because they are more vulnerable. Without the protection of cocoon shells, pupae were more likely to be bitten by flies and mosquitoes, and then they didn't moth but instead died. Spinning larvae after being bitten would also stop secreting silk.

Besides, the practice of the flat cocoon technique might also bring possible negative effects to silkworm's mental state, an empathetic assumption made by the chief biologist of Guangxi Sericulture Research Centre, Mr Zhang. When chatting in Zhang's office, he pulled open a drawer and showed me two samples of flat cocoons, one purely white and the other in vibrant yellow (Figure 3.3). The yellow silk was spun by a wild silkworm strain found in the autonomous county of the Yao ethnic group in Nandan, Guangxi province. The differences in colour and texture between the two silkworm varieties were much more conspicuous to the naked eye in the form of flat cocoons. As Mr Zhang put it:

> When silkworms were spinning on a flat plane, they would probably feel very uncomfortable and unsafe. Silkworms make cocoons to shield themselves, rather than produce silk. Their eyes are susceptible to light, and they move according to the layers' transmittance inside the cocoons to make sure the shell will eventually become evenly thick. Silkworms adjust their moving direction according to the light intensity within the cramped space inside the cocoon and then secrete silk filaments to seal up thinner areas and thus block the light as well as germs and natural enemies. When

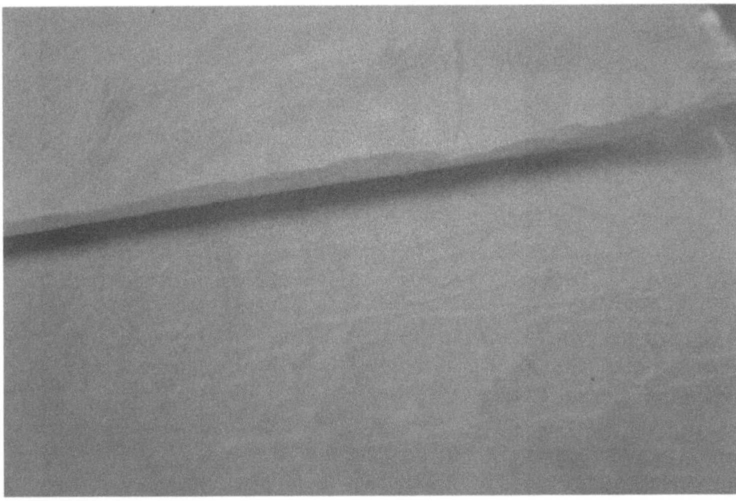

Figure 3.3 Flat cocoons in white and yellow in Zhang's lab. Photo by author

> spinning silk on the flat plane they must be unbearably distressed; they must think they failed to build a good and safe house to pupate.

In the scenarios of both silk quilt production and artistic creation, the application of the flat cocoon technique needs to be implemented in total darkness or a room with an evenly distributed light source. The ready-to-spin silkworms on a level plane will be deceived by the deliberately settled light conditions. There is neither any clue for them about the directions to move towards nor a fulcrum point to locate a three-dimensional space to create an ellipsoidal silk shell to reside. Zhang's anthropomorphic suspicion of the naked silkworm's mindset and emotional state cannot be scientifically verified but vividly points to the artificiality of light sources set according to the physiological characteristics of silkworms. The artist's conscious manipulation of the environment for the generation of flat cocoons is clearly contrary to silkworm habits, or natural instincts, a scene of intervention that could almost be considered the antithesis of the Daoist spirit advocating non-action and spontaneity. When the larvae had completed their mission of secreting silk in the flat-cocoon shape on various materials in Liang's art studio, they metamorphosed into pupae and were no longer of any use to the artist. Subsequently, they were left aside, some turning into moths and others dying for unknown reasons. Nonetheless, Liang had already realised that he was placing these insects at a higher risk than in an ellipsoidal cocoon, and hence asked biologists for advice on the possibility of lowering their chances of dying halfway.

Although this effort to reduce damage was temporarily futile, Liang was at least aware that his employment of the flat cocoon technique not only brought significant material transformation to silk threads in form but also forced silkworms to undergo discomfort at physical if not mental level.

Interlude: A (Failed) Trial with Hormones

The artificiality in Liang Shaoji's artistic experiments with the silkworm is foregrounded in his two encounters with hormones while raising silkworms and making art with the flat cocoon technique. For silkworm farmers, if the worms within their farmhouses were not all only healthy but also growing at a neat rhythm, namely, moulting, cocooning, and metamorphosing at almost the same pace, they then would be able to stop feeding with mulberry leaves at a specific point in time and then harvest the maximum amount of cocoons a few days later at another fixed moment. This temporal neatness saves time and labour as farmers won't need to bother waiting and caring for the silkworms that deviate from the speed of the majority. To meet the need for synchronicity in silkworm farming, silkworm biologists found hormone treatments as the antidote. There are two key signalling hormones in silkworms' metamorphosis from larval to pupal stages. Ecdysteroids accelerate the moulting speed of the silkworm, while juvenile hormones, known as JHs, maintain the insect's larval phase and their decline is essential for eliciting the transformation of a silkworm from larva to pupa (Hiruma and Kaneko 2013). This pair of hormones work on quite the opposite principles, and they are widely used in agricultural practice to regulate the developing tempo of silkworms.

The artist has his unique experience in tackling the silkworm hormones hands-on when he raised silkworms in rural Linhai in Zhejiang. Ecdysteroids are responsible for regulating the moulting process of an insect in its larval period, whose deficiency or excess might lead to abnormal development or even death. In sericultural practice, using this hormone helps increase the moulting pace of silkworm larva for upcoming metamorphosis into pupa; for a larva which is about to fully mature, an increased concentration of moulting hormones in its body will shorten the preparing time for its morphological transformation and thus it would be ready for spinning a cocoon more quickly. Ecdysteroids would not act by contact with the insect's body surface but only via ingestion, and thereby they are manufactured into a solution packed in a boxed vial and then sold to farmers as a common 'veterinary drug'. The instruction manual advises that when it is observed that some of the silkworms in the same batch are mature enough for cocooning, the farmer dilutes a vial of ecdysteroids solution with about 2.5l water for one sheet of worms (about 25,000) and sprays evenly on mulberry leaves before feeding them to the worms. Ideally, the silkworms falling behind will be accelerated to catch up with the majority, so that the whole batch will be prepared to climb up the cocooning frame to secrete silk in synchronised time after this hormone treatment.

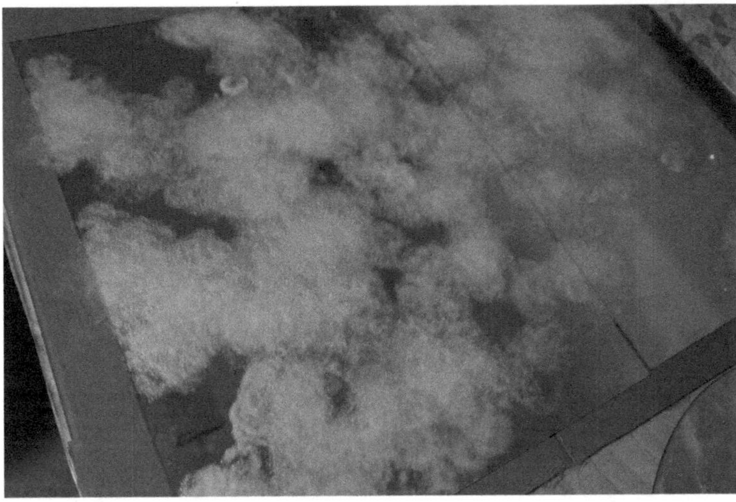

Figure 3.4 Silk traces in the figure of '8' on a mirror in Liang's studio. Photo by author

The artist's story with hormones started with an ecdysteroids overdose accident. In Liang's years of experiments on the flat cocoon technique and the suitability of different materials as surfaces, mirrors were among the tested objects. During my visit to Liang's studios at Tiantai, I saw several mirrors on the shelves covered with a thin and filmy layer of silk threads spun as a flat cocoon. Zooming in on those patterns of the silk pieces on each mirror, it was not hard to recognise that the traces of fibres are arranged in continuous figures of the number '8' overlapping each other (Figure 3.4). 'Figure-of-eight' is an accurate but rough behavioural description of silkworms' habits as well as a generalisation of the visual traits of silk traces secreted on flat planes. The size and width of 'eight' patterns, however, can subtly vary depending on individual silkworm's variety, size, or health condition. Liang not only recognised these nuanced variations but also noticed some exceptional deviant '8'. The first time Liang encountered an aberrant shape spun by a batch of silkworms was many years ago when he was staying in rural Linhai. As Liang recalled, he once fed silkworms with moulting hormones, the ecdysteroids, before they started to cocoon as a normal step in silkworm cultivation practice for a neater growing pace. That time he got the dose wrong quite by chance. Liang was surprised to detect that after the mistaken overdose of ecdysteroids, the amplitude of those spinning silkworms' head oscillation changed significantly, and as a result, the flat cocoon secreted on the mirror took on a much wider shape of '8'.

For farmers, an overdose of ecdysteroids would lead to a reduced income due to the thinner cocoon layers counting as poor-quality as the larvae were accelerated into the silk-secreting stage and did not manage to consume enough mulberry leaves prior to making cocoons. The quality and selling price of ellipsoidal cocoons, nonetheless, is not the artist's concern. Instead, out of an artist's sensitivity to visual elements, Liang was excited by this 'novel' shape of the figure eight which was triggered by his accidental application of hormone. Liang then decided to feed subsequent batches of silkworms with an overdose of ecdysteroids as an attempt to repeat the result of abnormal silk traces. However, to Liang's puzzlement, the awkward fatter shape of '8' resembling brass plates never reappeared despite his persistent trials.

Liang's second unexpected encounter with such strangely shaped silk traces is related to JHs which work in the opposite principle to the ecdysteroids on silkworms. The role of JHs is to maintain the silkworm in its larval stage from transforming to a pupa. At the end of a silkworm larva's fifth instar, JHs fade away in its body before cocooning and pupating. JH and its homologues have been widely employed in modern agriculture as a new generation of pesticides. Traditional pesticides applied in agricultural practice contain chlorine, arsenic, and organophosphates such as dichlorvos acting as nerve agents to kill unwanted insects in the field. Long-term exposure to these chemicals can easily result in poisoning to humans. New generations of pesticides are less toxic (to humans), and their chemical composition and structure have changed considerably with different working mechanisms. As one of the main categories of new-generation pesticides, hormones achieve the goal of insect control by affecting their energy metabolism and substance synthesis, JHs being the most used of this kind. When JHs are intentionally applied to insect larvae, they would pupate abnormally and die. Adult moths treated with JH pesticide would not be able to mate and lay eggs, whereas eggs laid by poisoned female moths will fail to hatch. For sericultural practitioners, however, JHs can be a fatal threat.

When I was talking with biologist Mr Zhang in Nanning on silkworm hormones, Zhang highlighted the fact that seventy percent of all pests worldwide in agricultural farming belong to the order lepidoptera, and unfortunately, the domesticated silkworm is one of the most comprehensively studied. The misfortune of silkworms lies in the indiscriminate attack of hormonal insecticides on the entire lepidoptera insects. The working mechanism of JHs applies to the whole lepidoptera order where they do not poison insects to death literally but prevent worms from undergoing metamorphosis and completing their life cycles. Silkworms in sericultural practice though are not the original target of hormonal pesticides but have frequently been the innocent victims of wider agricultural pest control measures. Zhang told me how this tragic misadventure occurred:

> In pest control, not only have pesticides been upgraded, but the way they are sprayed has also developed. In the past, farmers carried barrel-sized spray pumps on their backs to apply drugs while walking in the field, and later they drove sprinkler trucks around the farmland. More advanced types now spray from aircraft. Once the height has gone up enough, the

radius of applied pesticides becomes much wider, though at the expense of losing accurate focus on the targets. If there is a mulberry plantation near the farmland targeted for hormonal pest control, the mulberry leaves were likely to be contaminated with hormonal pesticides. As the concentration of JHs is not high, silkworms consuming the tainted leaves would not die immediately, but their larval stage ensues till the end of their lifetime, no longer cocooning or pupating anymore. Therefore, JH insecticide has turned into a major crisis for the sericultural industry even though they have helped the mankind in other agricultural practices.

The crisis caused by JH insecticide is so prominent in silkworm husbandry in China that even the artist Liang had come across such a mishap in his art-driven silkworm cultivation practice. It had been several years since Liang first encountered silk traces of the fatter figure of '8' shape. In 2018, he accidentally met the atypical eight-shaped again which he had failed to repeat by controlling ecdysteroids dosage in past experiments. Liang recalled the incident when we were talking in his studio in Tiantai:

> Last spring, I was raising silkworms in the countryside of Linhai. I was too old to cultivate many silkworms alone, so some local farmers came to help me pick mulberry leaves and feed the worms. Then there was one batch of silkworms that ate the leaves but did not spin cocoons at the expected point of time. The farmers noticed the phenomenon and were very anxious. One came to me ashamed and said: 'I'm so sorry Mr Liang, but it seems that all the silkworms are dying.' I didn't know what happened at that moment but informed the farmers not to dump the anomalous silkworms. I just couldn't bear to throw them away. I drove to the farmhouse immediately and stayed in the village for a whole day, picking out the silkworms that were still struggling to stay alive. I then took almost five hundred living silkworms back to my workshop in the neighbouring Tiantai county and placed them onto two pieces of mirror.
>
> I waited a few more days and some of the silkworms unexpectedly started spinning. More surprisingly, the silk traces spun on the mirrors shared the same wider shapes with those triggered by overdosed moulting hormones which I had been looking for over the years. Although the silkworms' heads still oscillated along a figure-of-eight course, the sway became noticeably larger, and they looked like a human suffering severe cramp. The overlapping strands of threads on the plate thereby formed circular silk pieces. Even weirder, these silkworms who finished spinning flat silk cocoons did not transform into moths eventually but stayed in their pupal stage till death.
>
> This time, a similar scene reappeared in front of my eyes, but I had no clue why it happened. To solve my confusion, I took photographs of those

> abnormal silkworms and the silk they spun on the mirrors and sent them to the experts at State Key Laboratory of Silkworm Genome Biology in Southwest University in Chongqing. Silkworm biologists there hadn't come across anything like this and couldn't give me an explanation. I then went back to ask the farmers how the mulberry leaves were picked. They remembered at that time the temperature was so high that some leaves in their own mulberry plantation had wilted and there was not enough food for their fifth-instar silkworms with a huge appetite. The farmers then turned to their neighbours for help and picked a few bags of leaves in their neighbours' farmlands nearby. I tracked it down and found out what went wrong. The land next to the neighbour's mulberry garden has peanuts grown there, and the peanut field had just been sprayed with pesticides during that period. Moreover, the drains in the peanut field were connected to the mulberry garden. So that batch of freak silkworms should have been poisoned by pesticides.

Liang didn't specify which kind of pesticide was the chief culprit, but by that time I was able to complete the final piece of this detective story puzzle on my own. 'Eating mulberry leaves all the time but not spinning cocoons' and 'forming cocoons without complete metamorphosis' are too obvious clues of an overdose of JHs. The symptoms presented to the biologists in Chongqing spotlighted the worms' wildly swinging heads and the round-shape pieces of silk they spun. Besides, the inability of silkworm pupae to complete the metamorphosis process after spinning flat cocoons is one of the negative consequences of implementing the flat cocoon technique even without the interference of hormones. These might be the reasons why experts failed to provide Liang a scientific answer. As for Liang's failure to repeat the targeted result by applying the ecdysteroids in experiments, I found a possible explanation in my conversation with biologist Zhang on the ecdysteroids' counterparts, the JHs. Besides functioning as pesticides, I also learnt from Zhang that JHs used to be applied intentionally to silkworms in sericultural trials. Precisely controlled doses of JHs can prolong the larval stage of a silkworm and delay its pupation as a means to increase the cocoon yield. When the larvae's fifth instar was extended two more days, as Zhang noted, the worms ate much more leaves and secreted thicker cocoons with 5% increased weight. The prolonged process of energy accumulation can convert more proteins in mulberry leaves into silk proteins. However, JHs as a hormonal treatment has never been promoted and used widely in silkworm cultivation, because the correct dose of JHs to be applied is not easy for farmers to learn and handle in practice. As Zhang explained, silkworms are particularly sensitive to hormones, and if the dose of JHs exceeded, there would be a risk that silkworm larvae simply would not cocoon anymore. In addition, a two days' delay of maturation to increase the silk yield by 5% means two more days' leaf-picking toil. Considering the cost of labour, applying JHs might not count as cost-effective and worth to try for most farmers. To achieve a specific outcome by using hormones without affecting the silkworms' metamorphosis usually calls for professionally trained skills and laboratory conditions. For ordinary people, a farmer or an artist like Liang, it would be difficult to precisely control the dose of JHs or the ecdysteroids for unconventional purposes.

Hormones have been intentionally experimented with by Liang to stimulate silkworms to secret silk filaments of dramatically abnormal shaped traces on two-dimensional platforms, which would be aesthetically innovative or attractive. Liang Shaoji's unique findings on silkworm hormones as an artist lie in his observation of the wilder and greater head swing amplitude of silk-secreting larvae reflected in the visual forms of silk traces they have spun. Such unusual phenomena were generated by two hormones with opposite principles of action, the ecdysteroids and the JHs. The artist's observations and (unsuccessful) experiments on how hormones affect silkworms and silk filaments have not been presented in artworks to the public. This 'backstage' episode, however, clearly demonstrates the intertwining relationship between making art and cultivating silkworms in his implementation of the flat cocoon technique, a dominant characteristic I have described as 'making-in-cultivating' in the previous chapter. Using hormones to interfere with the natural and spontaneous *'ziran'* biorhythms of silkworms is both a common technique in modern silkworm farming and an unexpected pathway for seeking artistic innovation.

In Liang's trials with silkworm hormones, the physical form of silk filaments was designed to undergo double 'rites of passage', namely, from the traditional ellipsoidal cocoon to the flat silk pieces and from the ordinary '8'-shaped flat cocoon to the abnormal round shape. At the same time, silkworms went through double-risk physiological challenges in the disorientation of secreting silk on a flat surface on the one hand and the pressure from hormonal intervention on the other. However, Liang's tactics are not more domineering compared to those of silkworm farmers. He placed silkworms on flat planes as silk quilt producers did and applied hormones as most silkworm feeders also practice. It was only serendipitous events that triggered Liang's exploration into the relationship between hormones and the shape of silk traces by blending these two means. Comparatively. Liang's determination and assertiveness in using the flat cocoon technique for artistic purposes are more prominent in his continuous experiments on materials as discussed below.

Experiments on Materials

Liang's silkworm art projects began in 1989 when he endeavoured to learn how to invite silkworms to spin silk on metal materials as his initial artistic vision, since for him metal represents 'the force of industry and rationality' (Liang 1994) as a strong contrast to soft organic material silk (Figure 3.5). In both historical and contemporary silkworm cultivation, farmers have been using lattice frames mostly made of paper, plastic, wooden, or bamboo strips

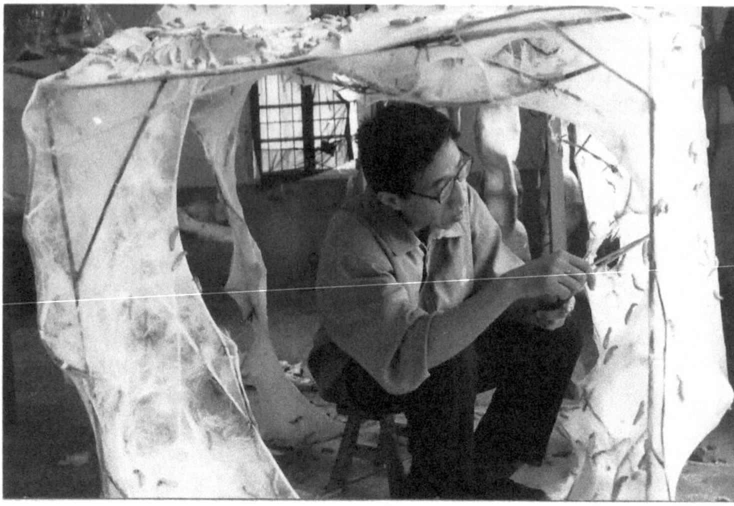

Figure 3.5 Silkworms and Liang Shaoji working together in the studio. Copyright: Liang Shaoji

for silkworm larvae to crawl and spin cocoons. Bales of straws were also commonly employed in some areas in China in the past. When the task for silkworms shifted from spinning ellipsoidal cocoons to secreting silk on flat surfaces, would they manage to accommodate alien materials such as metal, glass, and even the skin of human beings? This question drove Liang to explore the habitual and physiological characteristics of domestic silkworms.

In 1989, copper was the first material to be tested in Liang's experiments with the flat cocoon technique, as the artist preferred time-stamped materials such as rotten iron and rusty copper to stainless steel. However, when Liang cautiously placed a bunch of fifth-instar silkworm larvae onto some copper wires, they all moved away in a slow but determined way. Liang then sought advice from silkworm biologists at Zhejiang Agricultural University:

> I went to their labs and silkworm biologists there were very curious about my artistic project; but they had no ready answers whether silkworms have an innate preference for leaves, bamboo, wood and other natural things and dislike metallic materials. Experimenting with various surfaces for silkworm's spinning flat cocoons seemed far beyond the scope of scientific research and therefore those experts can offer little help. Sericultural study then was a production-driven discipline and merely aimed for a higher yield of cocoons, or a neat timing of silkworm's growth for convenient harvest and better quality of silk yarn. Biologists are concerned with uniformity, but I care more about the differences in materials and forms.

It could be imagined that a scientist working at an academic institution would hardly get a funding application approved for a proposed experiment that might fail to justify any prospective contribution to the silk industry, such as one examining whether silkworms prefer or not spinning flat cocoons on metals or glass. Silkworm biology does not cover Liang's artistic inquiry which shows little 'utility' in terms of improving efficiency or productivity in silkworm husbandry. Liang therefore decided to reside in a village in Linhai, Zhejiang and conducted his own experiment on silkworm's behavioural features by raising those creatures with local farmers. He made constant adjustments to lure silkworms into staying on the copper mesh for a longer period of time. Liang described to me where the difficulties lay in his approach to tackling the relationship between silkworms and surface materials: 'You should not grab the worms forcefully otherwise they won't secrete silk. Just let them move naturally as they wish. All I have to do is to create certain conditions for them to crawl'. To counter the resistance of silkworms to uncanny materials, Liang nevertheless still resorted to bioscience during the processes of his creative practice (see also in Chapter 2). One of Liang's strategies was to improve silkworms' stamina. In the lab of Linhai Silkworm Breeding Station, biologists assisted Liang to hybridise different silkworm varieties and Liang eventually selected a few promising 'candidate' breeds with stronger physical strength and higher immunity to continue his artistic experiment.

It took more than three years for Liang to observe an encouraging phenomenon: If the density of silkworms on a plane such as a piece of copper mesh was high enough, they would stay there longer and eventually spin some silk threads. This strategy sounds easy to implement. I, therefore, wondered whether it can be inferred that if one puts an extremely large number of silkworms on any flat surface, they will secrete flat silk flakes on it regardless of their tastes for the material. Liang explained that if the population were too dense, silkworms would feel uncomfortable. As the artist observed, should there be too many silkworms, there would not be sufficient room for each one to wave its head in a trajectory of '8' to secrete silk. Moreover, being squeezed by other silkworms in the crowd would also lead to breakage of their skin and thus higher susceptibility to pathogens.

Similar experiments were also conducted on human skin. Liang suspected that the smell of human hair and sweat is too pungent for silkworms when he did a 'pilot study' on his own body. Like the trial on copper mesh, Liang found out that if the density of silkworms on a plane is high enough, the odour of fellow worms, which is attractive for according to Liang's hypothesis, becomes stronger than the smell of the surface and thus a large group of silkworms would stay in a place unpleasant in smell for a longer time. Meanwhile, when it usually takes four to five days for silkworms to finish building their cocoons, Liang noticed that the worms are extremely effervescent and active on the first spinning day. As time slips by, silkworms become slower and less active, and the span of silkworms' body movements and head swing was much shorter in later silk-secreting days. These behavioural traits are both noticeable on metal and human skin. He explained how it felt when silkworms spun silk on his arm:

> Silkworms have a little lower body temperature than our humans and it feels cool when they lie on my arm. Too many silkworms spinning silk on your skin, however, can be painful, pa-pa-pa-pa, just like a lot of tiny needles sticking in your body. Silkworms might be in pain too; they hate the sweaty odour of your skin. Sweat protects human skin but smells monstrous for silkworms.

Since 1993, Liang has managed to make silk-wrapped installations where silkworms spin 'flat cocoons' to cover charred copper wires taken from burned-out engines ('Beds'), pyramid-shaped iron barbed wire ('Time and Permanence') and more challenging materials such as human skin and mirrors in performances such as 'Cloud Mirror' and 'Self-roped'. In Liang's observation of silkworms' behavioural patterns, silkworms prefer wood, plastic, and bamboo more than copper, iron, and glass. However, it is still unclear why such preferences exist. In Liang Shaoji's hypothesis, it is the silkworm's keen sense of smell as an insect that explains their different modes of action on a variety of materials as spinning workstations. Metal, human skin, and glass, for instance, must emit some odours imperceptible to humans but unpleasant for silkworms, while they are fonder of the aromas of leaves, bamboo, straw, and paper, which in Liang's words are more '*ziran*' and less artificial than man-made materials. Silkworms, therefore, have an affinity for natural materials.

Liang's assumption of silkworms' 'natural proximity to nature' is contested, which I will return to later, but this theory of olfactory sensation has strong explanatory power over the phenomenon of density. As Liang noted, when the number of silkworms within a certain area is so large that the smell of the silkworms' bodies overshadows the smell of the surface material itself, they can tolerate the annoying environment more than as a tiny flock. As for the silkworm's olfactory fondness, existing biological findings in silkworm fodder studies explore their smelling apparatus to improve more sensually attractive artificial food. Nonetheless, glass and metals are far beyond the category of edible raw materials for silkworm fodder formulas and therefore beyond the scientific scope of scrutiny. When visiting Guangxi Sericulture Research Centre in 2019, I asked about the correlation between silkworm and the material of silk-spinning surfaces, and there I unexpectedly heard another suspicion about the silkworm's repellence to metals. The biologist, Zhang, also observed with Liang that silkworms like wooden boards and dislike metallic materials, but his explanation focused on silkworm's tactile rather than olfactory sense:

> My colleague once conducted a comparative experiment between the glass and plastic plates in which silkworms seemed to lean towards the plastic.

> Their inclination might be ascribed to the material's friction. Silkworms have many pairs of crochets or hooked toes on their feet, and freshly spun silk fibres are coated with gum-like sericin to help them to crawl forwards. If the surface is particularly slippery, silkworms have to toil on with extra effort and suffer from a sense of insecurity. The silkworms would feel awkward when they were unable to move at will.

Zhang's 'theory' was narrated in an anthropomorphic tone and therefore might be persuasive but this 'tactile' interpretation of silkworms' liking of certain materials, however, seemed slightly unhelpful when it comes to the population's density phenomenon that Liang observed. Whether tactile or olfactory, both Zhang and Liang's attempts to understand silkworm's behaviour patterns pay close attention to their sensory experience with considerable imagination and empathy. Those two approaches might not be mutually exclusive. They could be compatible to illuminate the 'presumed' sufferings of the worms on metal surfaces where silkworms would be enduring the eerie stink and simultaneously the strenuous crawl when being placed on an iron or copper board to spin flat cocoons.

In Liang Shaoji's rough classification, wood, bamboo, and stone are '*ziran*' (natural) as they are associated with plants and earth, while 'unnatural' copper, iron, and glass are intensively used in manufacturing industry. This division correlates strongly with the likes and dislikes of silkworms. The Chinese term '*ziran*', however, is a nuanced and ambiguous category in terms of describing materials. Some metal and glass materials, for example, appear abundantly in soil and minerals, neither synthetic nor artificial, and therefore might also be nominated as 'natural' just like the stone. There are also counterexamples in Liang's artistic experiments where human skin, for instance, is organic but not favoured by silkworms, whereas man-made plastic tools function well in both silkworm husbandry and Liang's works. The naturalness of materials as the workstation for implementing the flat cocoon technique is therefore an untenable pseudo-proposition in a practical sense. However, it is arguable '*ziran*' as an unstable classification criterion for surface materials may yet still be significant on a symbolic level, a point I will return to in the next chapter in relation to specific artworks. Liang, himself, reckons that his silkworm art series is executing the principle '*tian-ren-he-yi*' (天人合一) in Chinese Daoist philosophy, 'the integrity or oneness of human and heaven/nature'. Therefore, he had not been disappointed by the numerous 'failures' during his experiments on silkworms and various materials before he could finally realise some projects as designed:

> I just intended to make art happen. I wondered whether the idea of putting silkworms on glass or metal would go against the Daoist philosophy of harmonious symbiosis. If so, I won't deem it to be a failure, but another artistic outcome, an alternative answer to my question on '*tian-ren-he-yi*'.

In this careful examination of technical details of inviting silkworms to spin silk on flat surfaces of various materials, their uncomfortable bodily experience has been foregrounded, which is thoroughly neglected in critics' ethical evaluation of Liang Shaoji's artistic approach. As noted in the previous chapter, Liang's artistic practice encompasses the cultivation of silkworms and making art and therefore it will be helpful to juxtapose his deeds with silkworm cultivation for pure profiting purposes. The explicit discussions of Liang's techniques above provide more accurate and comprehensive analytical materials for further pondering on the ethics, or at least Daoist spirituality in his art practice.

Rethinking Intervention and Naturalness

Regarding the naturalness in the processes of flat silk generation in Liang's artworks, Yang (2018: 125–126) comments that the artist's use of silkworms is an interaction in accordance with the principles of '*ziran*', reflecting the Daoist understanding of the relationship between humankind and nature. Daoism, as Yang argues, opposes forceful actions that go against the ways things naturally go and dissolves the hierarchical distinctions among species which has fostered Liang's compassion, appreciation, and respect for the silkworm (2018). However, as demonstrated in this chapter, Liang's artistic practice is far from the Eastern idyllic and non-interventionist way of dealing with insects as Yang contends.

Intensive silkworm cultivation for silk industry in China can be intrinsically manipulative especially in contemporary society with farmers' more mature experience in conducting hormone treatment. Most silkworms that spin ellipsoidal cocoons are killed in the pupal stage within the cocoons in silk reeling factories while being processed in drying or hot water reeling machines. In small-scale farms where the silkworms spin flat cocoons for silk quilt production, they are allowed to pupate and moth instead of being boiled to obtain silk, but their health is compromised without the protection of cocoons and thus they are more likely to die straight after secreting flat silk than to complete their metamorphosis. Farmers, moreover, have used the application of hormones and the flat cocoon technique long before Liang's experiments. These methods to harness the biological instincts of silkworms as domestic species that are highly dependent on human care for their survival and reproduction. The artist's innovation in practice lies in the introduction of many uncommon materials into the application of the flat cocoon technique. Such uncanny materials such as rusty copper wire mesh, mirrors, and human skin are likely to aggravate the existing discomfort of silkworms secreting silk on flat planes, either because of materials are too slippery or too pungent for the worms.

The social relationship between humans and the domestic silkworm, mediated by silk, has been carried out using a human-centric exploitative and manipulative approach for thousands of years. When the survival of this vulnerable insect has a one-directional dependency on humans' care, the inequality in silkworm-human power relations is magnified. Regarding the Daoist principle of 'following the way it intrinsically goes', silkworm husbandry as an industry can be seen as full of artificiality and unnaturalness. On the other hand, the flat cocoon technique, either implemented by farmers or Liang Shaoji, makes use of the biological traits of the silkworm rather than working against the 'natural laws'. In this sense, Liang's making-in-cultivating art practice sits at the liminal stage of being and not being *'ziran'*. The techniques employed by Liang are less cruel compared to boiling silkworm pupae within the cocoons in silk reeling factories but cause more possible uneasiness to silkworms placed on unfamiliar materials. Liang's experiments on materiality and the flat cocoon technique to realise his own artistic vision is surely manipulative but not necessarily 'un-natural'. He had set up more difficult conditions for the silkworms to endure and adapt for their own growth and for the generation of silk.

When the concept of '*ziran*' implies integrity as a whole and harmony, the Daoist inaction '*wuwei*' shouldn't be a refusal of contact or mutual isolation. Liang defended his artistic experiments during my interview in his studio: 'I want to open up the neglected potential of the silkworm and see whether a kind of harmony can be achieved'. Though the silkworms' behaviours significantly demonstrated their simultaneous aversion to certain materials, Liang was determined to create encounters between them rather than retreating to conservative preservationism. Those generated encounters are not necessarily romantic or pleasant but are likely accompanied by hurt, anxiety, and insecurity. The key verb '*he*' (unity, harmony, togetherness) in '*tian-ren-he-yi*' emphasises the process to meet, join, confront, and unite, as myriad things have already been separated in various ways in the Daoist cosmology. However, getting along with other living beings requires wit and courage. Liang's art-making and silkworm-cultivating scenarios hence show the intra-contradiction of the concept '*ziran*'.

Liang's 'making-in-cultivating' techniques involve various forms of manipulation, intervention and control rooted in intense cultivating labour and bioengineering, which are not always consistent with Daoist advocacy of inaction, naturalness, and spontaneity in face of nonhumans. In Liang's artworks, flat silk flakes are thus semi-natural semi-artificial material, rather than signifying pure, primaeval, and intact nature uninfluenced by progressivism and industrialisation. The complicated growing processes of flat silk in unnatural situations for silkworms thus open up the interpretive potentials of silk in Liang's artworks, which should not be simply designated as a semiotic-material sign of the natural and organic, as discussed in the next chapter.

References

Hiruma, K. and Kaneko, Y., 2013. Hormonal Regulation of Insect Metamorphosis with Special Reference to Juvenile Hormone Biosynthesis. *Current Topics in Developmental Biology*, 103, 73–100.

Liang, S., 1994. About the Nature Series [online]. *ShanghART Gallery*. Available from: https://www.shanghartgallery.com/galleryarchive/texts/id/54.

von Drathen, D., 2011. Liang Shaoji: Of the Immaterial and Its Corporeality. *In*: M. Bloemheuvel and T. van Kooten, eds. *Windflower: Perceptions of Nature*. Rotterdam: NAi Publishers.

Xia, K., 2015. The Silk Shadow of Chora-topia [online]. *ShanghART Gallery*. Available from: https://www.shanghartgallery.com/galleryarchive/texts/id/8092.

Xia, K., 2017. Kiefer and Liang Shaoji: Differences of Modernity between East and West in Integrating Three Worlds. *Northern Art*, 12, 60–61.

Yang, J., 2018. Dao of the Silkworm: The Ecological Thrust of Liang Shaoji's Nature Series. *New Arts: Journal of the National Academy of Art*, 39 (6), 119–127.

Zhao, F., 2005. *A Comprehensive History of Silk* (中国丝绸通史). Suzhou: Suzhou University Press.

Zulueta, C.C., 2016. Two Artists, Two Ecologies, and a Shared Empathy Towards Non-Human Animals' Agencies: Yanagi Yukinori and Liang Shaoji. *Journal of Contemporary Chinese Art*, 3 (3), 377–387.

4 Agency in Silkworm Rituals

'Planar Tunnel' (2011–2012) (Figure 4.1) stands unique in Liang Shaoji's oeuvre where silk floss generated by the flat cocoon technique exists independently, without interacting with other materials. When displayed on white gallery walls, these extremely thin flat silk pieces reveal an almost ethereal quality – reflecting jade-like or metallic sheens under varying light sources, with shadows duplicating their delicate textures on the wall. So gossamer-light are these silk planes that even a viewer's breath can cause them to tremble.

If viewed as a minimalist painting, its canvas is the painting itself, being extremely abstract and therefore highly versatile for symbolic interpretation. Art critic Xia (2015), for instance, interprets the circular silk shapes as echoing ancient Chinese jade pendants, suggesting a cosmological representation

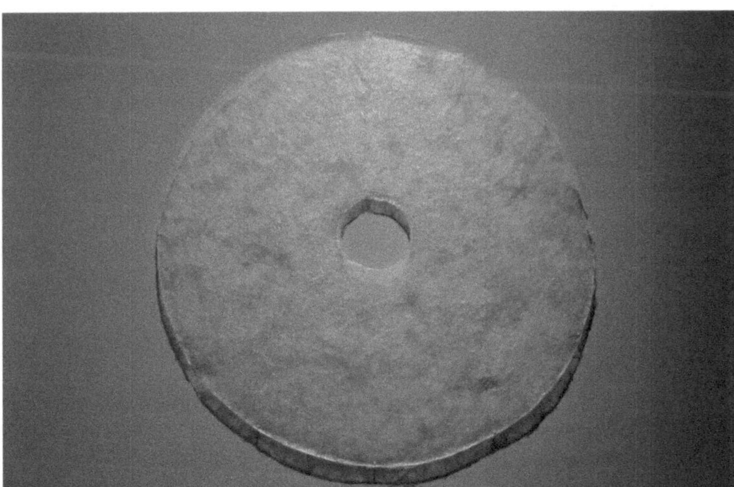

Figure 4.1 'Planar Tunnel', 2011–2012. Photo by author

DOI: 10.4324/9781003569091-4

encapsulated in the phrase 'heaven round and earth square' (天圆地方), and moreover, regards the piece as transforming a 'natural thing' – silk spun by silkworms – into art while preserving its inherent 'naturalness', embodying an Eastern aesthetic of ethereality and ambiguity. However, if drawing from the intricate interplay between cultivating silkworms and implementing the flat cocoon technique explored in Chapter 3, 'Planar Tunnel' cannot be simplistically equated with a pure and uncontaminated nature, or naturalness in silk, transformed or converted to art. Instead, it represents a nuanced compromise between the artist's mastery and silkworms' capacities, and hence the intrinsic paradoxes of '*ziran*' regarding the 'natural' form of silk and the 'natural' behaviour of the silkworm. Liang envisioned the silk's fundamental characteristics – its size, shape, and thickness – while silkworms realised their designs through the texture and lustre of each delicate flake. This process is much closer to a creative ritual of 'growing' that explores the concept of '*ziran*' both materially and symbolically.

The artist acknowledges that the Chinese material culture of silk has nourished his artwork, although he explicitly emphasises his stance of keeping a distance from the current mainstream ideology-infused Silk Road discourse appropriated by the Chinese Communist Party for diplomatic and cultural propaganda in present days (in Gladston 2011). His inspiration for 'Planar Tunnel', as he clearly stated, connects to a remarkable artefact: A plaint gauzy silk gown unearthed from Han Tomb No.1 at Mawangdui of Changsha City, dating to the Western Han Dynasty (206 BC – 25 AD). This extraordinarily preserved garment – now stored in Hunan Museum and weighing merely 49 grams across 2.6 square meters, as one of the earliest and thinnest dresses in the world – shares a profound aesthetic resonance with 'Planar Tunnel', both celebrating the sublime lightness, thinness, and translucency of silk fibres.

Meanwhile, in Liang's other numerous works, silk floss is no longer isolated but spun on objects of diverse materials. How flat cocoons meet and interact with these objects can reveal the silk's and even silkworms' functional and symbolic significance in his semi-ritualistic art – a proposed analytical framework this chapter will adopt. To better understand the compositional 'wrapped in silk' structure permeating most of Liang's projects, I will first allude to prehistorical rites involving silk fabrics and archaeological findings on silk and silkworm cultivation in ancient China.

Silk in Chinese Culture

Though my ritualistic framework anchors in ancient Chinese contexts, studies on silk and silkworms span a fascinating breadth of disciplines beyond a single cultural or temporal boundary. Scientists have endeavoured to understand the biological traits and potentials of the silkworm, the mulberry tree, and the silk (e.g. Hiruma and Kaneko 2013; Wang et al 2016). In humanities, topics such as the gender division of labour in ancient silkworm-raising practice (Chen

2013), the relationship between the legend of Silkworm Gods and Buddhism (Young 2013, 2017), and the iconology of sericultural handbooks (Hammers 2011) have drawn interest across disciplines. Silk Road studies, moreover, form a distinct body of scholarship that explores the socio-economic and cultural histories around silk production and circulation within global trade networks (e.g. Hansen 2015; Rong 2003; Yin 2007). Contemporary sericultural practices in regions outside China have also received attention, showing the diverse social functions and meanings of the silk-silkworm assemblage in different socio-cultural contexts. Onaga (2013), for instance, focuses on the tension among race, gender, and bioscience in Japanese silkworm sciences, while Vallard (2013) investigates wild silkworm species in textile industries in Thailand and Laos. Douny's anthropological study (2017) zooms in on wild silk indigo textiles in relation to indigenous cosmologies across West Africa.

Within the China context, Zhao's encyclopaedic 'A Comprehensive History of Silk' (2005) meticulously chronicles the technological evolution and knowledge updates of silkworm raising and silk processing methods across dynasties, as well as the artistic styles and artisan skills associated with silk fabrication. Bray (1997) narrows her focus to the late Ming dynasty (1000–1800), exploring the gendered division of labour in silkworm husbandry and commercialised silk-reeling factories through a feminist lens. Stockard (1989, 2002) also addresses gender and labour issues, conducting comparative fieldwork in silk-reeling factories in Shunde County, China, and Mansfield, Connecticut, USA. In these studies, silk and silkworms often serve as entry points to address broader themes within human societies, including economic, cultural, and political dynamics.

However, silk in ancient China was not merely a material commodity or a subject of labour; it held profound ritual and cosmological significance. Ancient Chinese ritual beliefs divided the universe into three layers – deity heaven, human earth, and the netherworld (Gao 1994 in Fung 2013: 221). Silk was no ordinary fabric but a sacred medium facilitating communication between the human and divine realms, as evidenced by archaeological discoveries and historical records. For instance, silk fragments from urn burials in the Yangshao culture of Qingtai Village, Henan Province, dating back around 5,000 years, reveal its role in funerary practices.

In this culture where nearly half of the entombments were in urn/jar burials, children's bodies were wrapped in silk before being placed in clay jars as coffins and buried near their houses (Fan 2020: 19). Clay containers were used for dead children because the ancients believed the close contact of the burial pot made of soil could grant the deceased the blessings from the goddess of the land. Moreover, the pottery was also a symbolic and material resemblance of the womb of a mother for the dead's rebirth, as speculated based on sympathetic magic and transmission theories (Lu 2001; Poo 1998). The jars, often left with a small hole, were believed to allow the soul to ascend to heaven, akin to metamorphosed silkworm moths emerging from cocoons (Zhang and

Gao 1999: 3; Zhao 2012: 15). In Zhao's (2012: 19) view, it was also because the silk covers the deceased like the cocoon protects the silkworm pupa that 'wrapped in silk' thus acquired its function of rebirth and transcendence in ancient Chinese obsequies. This ritual pattern also appeared in the Yin Shang period where bronze vessels as '*liqi*' were buried wrapped in silk. '*Liqi*' refers to shamanic objects to communicate respectfully to the supernatural in funeral ceremonies including vessels, paintings, jewellery, and other artefacts carrying cultural symbols in their shape, colour, and decoration (Fung 2013: 221). Silk was not a supplement to pottery jars and bronze vessels as '*liqi*' in rituals, but silk fabric itself can serve as a medium to 'bid farewell to the dead' (*songsi*) and 'communicate with the gods' (*shi guishen*), as recorded in '*Liji*' (Book of Rites), a collection of ritual rules written during the 5th century BC to the former Han periods. Zhang Hua (232–300) in his short story collection '*Bowuzhi*' (Vast Records about Different Topics) made association between the two homophones '*yong*', namely, silkworm pupa (蛹) and tomb figure (俑) to justify the ritual significance of silk as a mediating material. These ritualistic origins imbued silk with a spiritual dimension long before it became a fabric of daily life after the Warring States period (fifth century to 221 BCE), marking its enduring significance in Chinese material culture (Zhao 2005).

'Wrapped in Silk' as a Ritual Pattern

As a complete metamorphic creature, the domestic silkworm undergoes drastic morphological transformations during its weeks-long lifespan. The silkworm's ability to change body forms and secrete cocoons imbues both the creature and the silk it produces with rich functional potential, both materially and symbolically, particularly in funerary contexts. By highlighting the inferential agency of silk as a shamanic mediator in ancient Chinese funerary ceremonies, I draw a comparison between these prehistoric ritual practices and Liang Shaoji's contemporary silkworm-based artworks. In primaeval death rites, silk fabrics were reeled and woven from ellipsoidal cocoons, whereas in Liang's studio, silkworms directly secreted flat silk on a pre-arranged range of objects. By viewing the making of Liang's installations as a ritual process, silk became part of the artistic rite from the very beginning, rather than waiting to be degummed and woven into fabric. Acting as a mediator between silkworms and other materials, the artist facilitated a process in which silkworms were placed to crawl on various objects, spinning flat cocoons to envelop them. It was the silkworms, rather than the human artist, that executed the acts of covering, wrapping, and shielding – much like shamans.

In calamity-themed installations such as 'Helmets' (2004), 'Wenchuan Stones' (2014–2016), and 'Destiny' (2012–2014) (Figure 4.2), silkworms had close physical contact with helmets and headlights of deceased miner, as well as the rubble from the ruins of the earthquake and iron plates resembling the wreckage of sunken ships in the Gulf War, crawling on and cocooning over

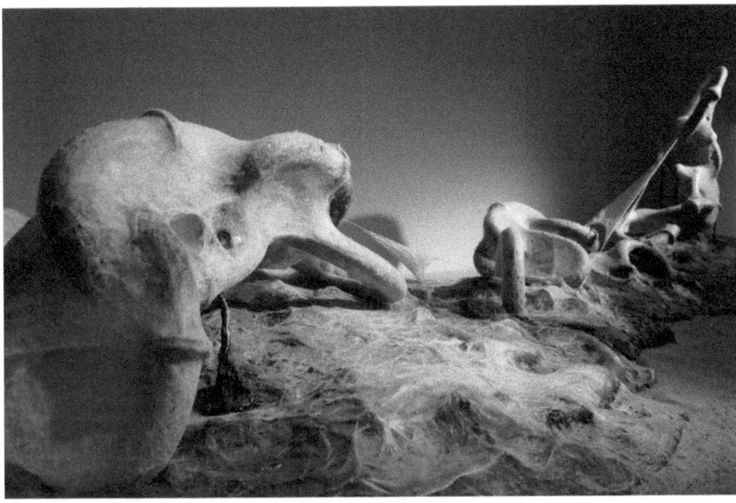

Figure 4.2 'Destiny', 2012–2014. Photo by author

those objects. The antique ritualistic structure 'wrapped in silk' thus has been revived in Liang's silkworm art while Liang's implementation of the flat cocoon technique involved the silkworm becoming an active initiator of the healing and soul-releasing rituals. The visual presentation of 'wrapped in flat silk' where the contours of objects are 'smothered over by the silk floss', however, has been read by von Drathen (2011) as softening and receding the dread in order to obstruct attempts to preserve memories of horror, which can be a potential covering-up by the government. This, however, is a misinterpretation if considering the efficacy of silk in communicating with the supernatural and ferrying the souls of the dead to the otherworld in ancient Chinese ritual beliefs. On the other hand, Zulueta rightly points to the significance of the wrapping process before installations were exhibited in gallery space which '[extend] the protective cushion of life to enfold new forms of vulnerability' to mourn the death (2016: 384), though he does not elaborate on the wrapping process itself and neglects the ritual function of silk contextualised in Chinese culture.

The healing and repairing ritual in 'Snow Cover' (2013–2014) is less explicitly expressed compared to those directly dealing with disasters and death, even though Liang has pointed out on a symbolic level that the intention of this work is 'to heal and salvage', as mentioned in the prologue. The ordering in this artwork, namely, the discarded everyday objects scattered on the table, conveys a very clear iconographic connotation to the audience, representing ordinary fragmented life in contemporary industrialised societies (Xia 2017: 61). However, the ritual process to regulate the social dysfunction brought about by modernity is yet under-analysed. The agency of silk in this work is manifested in its capacity not only to cover and heal humans' wounds from dysfunctional social relations but also to animate and recover the abandoned non-organic objects whose social lives in human society have ended. 'Snow Cover' therefore should not be understood as an ironic presentation contrasting the illness of contemporary social life with the comforting power of vibrant wild nature as the silk denotes as Xia (2015, 2017) contends. Rather, this artwork initiates a ritual to reconcile complex social networks among humans, things, and nonhuman organisms in the contemporary world, via the intimate entanglement between man-made things, spinning silkworms, and the generated silk in the contact zone of silkworms and those objects.

On the other hand, the often-ignored temporality in Liang's ritual art with the implementation of the flat cocoon technique has been foregrounded in the ritual of ever-growing 'Time and Permanence' from 1992 to today, where batches of silkworms secreted silk to wrap not only the pyramid-shaped barbed iron wires but also the slightly duller silk floss spun by their silkworm peers in the past years. The newly spun silk threads forming the outer layers are relatively white and shimmering while the inner fibres gradually lose their gloss over time. This notable gradation of visual characteristics of silk floss on the same barbed iron pyramid shows the growth, or material accumulation of this artwork over nearly 20 years and commemorates the death of countless short-lived silkworms who had spent their whole spinning days on these objects. The conspicuously contrast in the temporal scale between the lifespans of silk and silkworm is exposed in 'Time and Permanence' as well as other Liang's installations using flat silk flakes. The short lifespan of numerous silkworms is thus superimposed on the temporality of Liang's processual ritual art, a key point to understanding his works that are often obscured by presenting the silk-bounded installations as 'end products' in exhibition settings.

Silkworms as Agentive Performers

The agency of silkworms becomes more prominent in Liang's two performative artworks, namely, 'Self-roped' (2000) and 'Cloud Mirror' (2007), in both of which silkworms were still required to spin flat silk floss over materials unpleasant to them. However, silkworms no longer only served as the producer of the flat silk floss but entered the processes of Liang's ritual art as main performers. 'Cloud Mirror' (2007) is a performative project (Figure 4.3) where Liang Shaoji, silkworms, and the silk floss all participated in this meditation ritual. The performance was situated on the summit of Tiantai Mountain where Guoqing Temple, the ancestral palace of Tiantai school in Buddhism is located (Figure 4.4). The artist sat cross-legged on a stone with a piece of glass mirror on his knee. Stronger crossbred silkworms could be seen crawling and spinning on the mirror, whose surface was then gradually covered by silk threads with the reflection of flowing clouds up in the sky. The artist kept still and meditated, while the tiny creatures on the

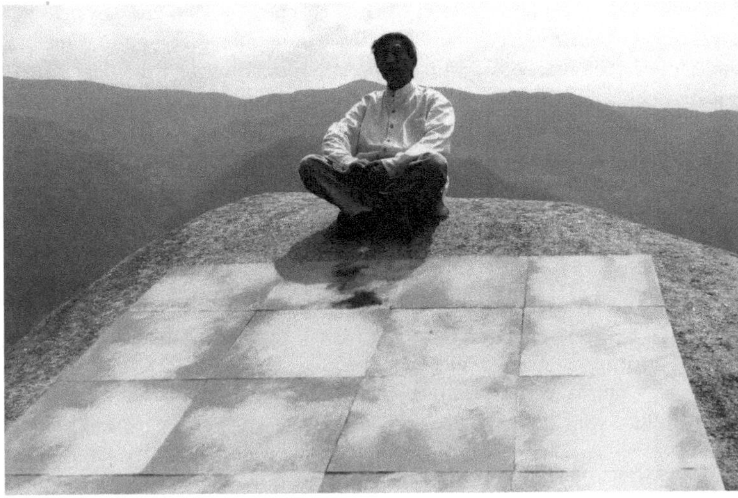

Figure 4.3 'Cloud Mirror/Nature Series No. 101', 2007. Copyright: Liang Shaoji

mirror strived to move forward and secreted silk at a slow pace. The material assemblage of silk and mirror, the surrounding mountain peaks, sky, cloud, daylight, and the underlying local Buddhist and Daoist culture and history of Tiantai were all compressed into this ritual event. The scene was visually idyllic and ethereal to contemplate, as manifested in archival photos.

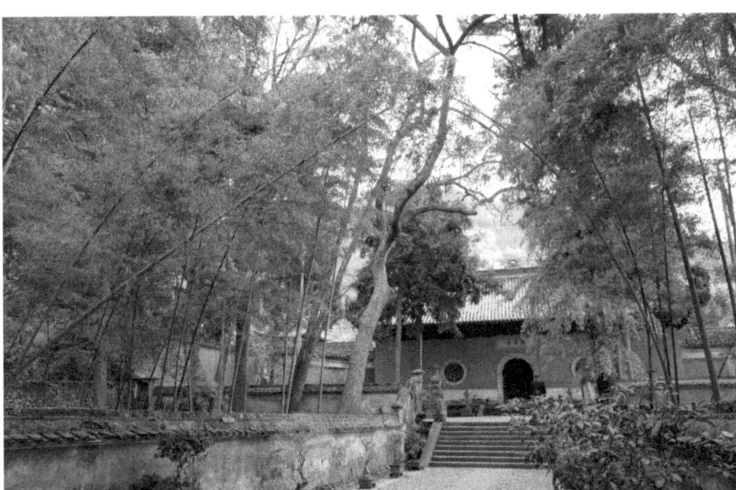

Figure 4.4 Guoqing Temple at Tiantai Mountain. Photo by author

Under the spotlight of 'Cloud Mirror' is the artist's on-site physical and psychological experience, a human being's immersion and spiritual meditation within a spontaneous spatio-temporal realm. However, the dark side of this artwork's celestial, harmonious, and intelligent presentation of '*ziran*' is the physical and psychological experience of the silent non-human participants. Dwelling and spinning on the mirror must have already drained the silkworms completely because spinning flat silk on slippery glass surfaces would bring double discomfort to the worms, as explained in Chapter 3. Their likely sense of fatigue, anxiety, and insecurity dramatically contradicts the calm, speculative, and serene state of the artist Liang in the same meditation ritual. Measured from an anthropocentric perspective, this ritual successfully conveys the spirituality of '*ziran*', or nature, despite the term's ambiguity, as the performance intentionally combined elements namely, the mountain, clouds, sky, animal, and cloud-like silk floss conventionally designated as being 'natural'. Nonetheless, if taking the silkworm's physical experience in this meditative ritual into consideration, the efficacy of this performance is rather undermined by the contradictory 'unnatural' state of the worms.

The uncomfortable experience of the silkworms is amplified and complicated in another performative work of Liang Shaoji. 'Self-roped' was performed indoors at the 5th Biennale d'Art Contemporain de Lyon in 2000, as part of the group exhibition entitled '*Partage d'Exotismes*' (Sharing of Exoticisms). Lyon used to be renowned for its well-developed silk industry (Hafter 2020: 33). Liang started to raise silkworms in the city before the opening of the biennale and invited two female students from a local dance school in Lyon to participate in this project. During the exhibition, these two women sat and lay at a corner of the gallery, naked and almost motionless, while hundreds of silkworms were crawling on their bodies constantly spinning silk threads to wrap them up throughout the daytime. Gradually, the dancers were encased by the silkworms using their self-spun silk 'ropes' (Figure 4.5). The allusion of the work's Chinese name 'Self-roped', or '*zifu*' (自缚), is a shortened version of a common idiom known as 'make cocoons to tie up yourself' (作茧自缚). The literal meaning of this idiom would be neutrally descriptive, denoting the pupation phase when a silkworm larva makes a cocoon to wrap itself until metamorphosing into a mature moth. Being covered up by cocoons, silkworm pupae are restricted to a very tiny room for movement while they are also protected from pathogens and natural enemies outside their ellipsoid-shaped shelter. Nevertheless, when the phrase 'make cocoons to tie up yourself' is applied metaphorically in Chinese, the restrictive aspect of a cocoon's function has been emphasised and thus the idiom usually indicates 'being trapped by self-imposed limits' or simply 'making trouble for oneself'.

The title 'Self-roped' fits ironically in this performance. When the silkworms had to spin 'flat cocoons' on human skin rather than an ellipsoid one for self-protection, the role of newly spun silk threads as a shield dissolved. Flat cocoons can no longer safeguard the worms, nor can they keep two female

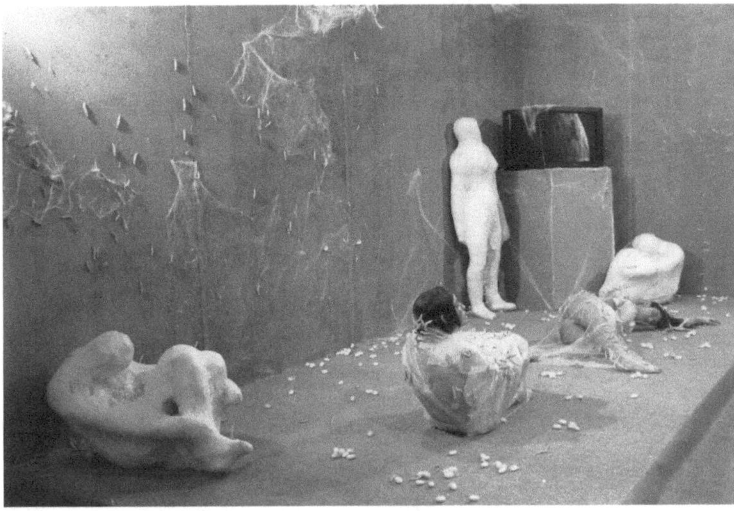

Figure 4.5 'Self-roped/Nature Series No. 31', 2000. Copyright: Liang Shaoji

dancers from any harm. Two roped human beings, at the same time, were constrained by the silk threads from moving freely. It seemed to be a lose-lose situation when humans intervened in the self-roping process of the silkworms. To implement this performance, silkworms had to tolerate the pungent smell of human hair and sweat according to the artist's empathetic speculation based on his observation of the worms' behavioural traits. The artist therefore had to increase the density of the silkworms on human skin to lure the worms to stay and secrete silk there for a longer period of time (see Chapter 3). In Lyon, Liang needed to make extra effort to facilitate this artistic ritual, as the design of the performative strategy was unfriendly to both silkworms and the two dancers. The two women not only endured the discomfort on their skin when the silkworms crawled on them but also suffered numbness from sitting in the gallery space for hours without any dramatic movement. They were even advised to eat and drink less so that during the exhibition time they would not need to go to the toilet throughout the whole day's performance. As for the silkworms, the artist observed in the pilot study where silkworms secreted silk on his own skin that silkworms are less energetic and active in the later phase of their four- or five-days' long silk-secreting period. For this reason, Liang carefully considered the growing pace of the worms and deliberately chose larvae in their second or third spinning day to perform on-site with their human co-performers. The silkworms placed on the dancers' skin were hence more docile and cooperative. In the end, both parties in this performance, as Liang noted, were 'too weary and almost lost all their alertness'.

For viewers and reviewers of 'Self-roped', it was not evident that the realisation of this performance requires strong perseverance from both silkworms and human beings. The scene appeared to be harmonious and serene, where two species cohabited intimately within a limited space and time. The art critic Cheng (2007: 68), for example, perceived the scene as silkworms caressing and protecting the female dancers as if they were silkworm pupae to metamorphosise. However, she also criticised Liang's patriarchal treatment of female models as a male artist in which Liang implicitly makes an analogy between the status of compliant women and of domesticated silkworms as 'soft, docile, industrious and impressionable' (2007). Cheng's feminist critique is incisive. If the human performers of this ritual had been males instead, it might not have affected the expression and efficacy of this performance, but the reasons for Liang's choice of female dancers were unstated. However, knowing of the unspoken endurance and exhaustion behind the scenes, Cheng's description of the cross-species interaction in 'Self-roped' seems overly romanticised. This performance would not be 'successful' should any of the human or non-human performers stay very sensitive and clear-headed. It was the enduring numbness that had triumphed over their physical discomfort and then guaranteed the transient but challenging interaction in this ritual. In critical discussions of 'staying with' other species, notions of care, affect, and attention are often emphasised and expounded (cf. Haraway 2008) while the strategic endurance, exhaustion, and numbness are too easily underestimated. The case of 'Self-roped' points to another possible gateway to meeting with other animals. Mutual tolerance or even ignorance can be reached via being tough, hardy, dull, and less egocentric. Living creatures involved in a difficult situation might need to employ this insensitivity towards each other in order to refrain from certain conflict or repellence. When human bodies and silkworm bodies came into direct contact in 'Self-roped', they were pushed to their physical limits and thus created a confrontational tension through the medium of silk. This ritual performance thus can be interpreted as an anti-structural (cf. Turner 1982) recognition of '*ziran*' in its initiation of an unnatural co-existence. The function of silk filaments sat somewhere between (unwanted) protection and confinement for both humans and silkworms, while the bodies in soft confinements were faced with the dilemma between freedom and security. As the artist states, the initial vision of this performative art originated in 1992 (Liang 2002: 43), not long after the Tiananmen crackdown. This work might also be Liang's reflection upon the unlimited state power and individual freedom, though in a veiled way.

The Efficacy and Presentation of Ritual Art

Among Liang's flat cocoon artworks, whether classified as installations or performances, the growing and generating processes enacted by three key agents – the silk, the silkworms, and the artist – determine the works' expressive content and ritual efficacy. As previously noted, the flat silk floss secreted in the physical contact zones between silkworms and various materials differs significantly from the ellipsoidal and flat cocoons produced in agricultural contexts. Therefore, understanding the agency of silk and silkworms in Liang's ritual art projects requires moving beyond the conventional stereotypes in Chinese material culture, where silk signifies rarity and preciousness, and silkworms are idealised as diligent and selfless labourers. Contextualising these projects in Chinese ancient funeral rites when silk had not yet been applied as daily clothing fabric evades the contemporary Silk Road discourse contaminated and appropriated by Chinese Communist Party's nationalist ideology which the artist too intentionally avoids. Viewing Liang's works through the lens of ritual shifts the focus from their final presentation in galleries to the artist's 'cultivating-in-making' practice, intricately intertwined with the silkworms' spinning processes.

The ritual structure of 'wrapped in silk' across Liang's works evokes ancient funeral rites where silk played a mediating role in healing wounds, repairing cracks, communicating with the supernatural, and ferrying the deceased to the otherworld. Such ritualistic functions of silk are tied to its physical and symbolic qualities: The cocoon's ellipsoidal shape as a protective shelter, the soft texture of silk, the dramatic transformations during the silkworm's metamorphosis, and the violent emergence of the moth breaking free from its cocoon. Liang's use of the flat cocoon technique adds complexity to this interplay. Silkworms are no longer passive silk producers behind the scenes but active participants in constructing meaning and enacting the rituals. In Liang's art, they assume the role of human shamans in ritual practice, crawling over objects and human bodies, producing silk as both a spiritual and material medium. This process invokes supernatural power in meditative or healing rituals, connecting Liang's contemporary practice to ancient Chinese traditions. Liang's silkworm art, therefore, can be seen as a microcosm of an animistic world imbued with myriads of energies.

This analytical framework, however, also raises critical questions about how such art projects – with their layered temporalities and active agencies – are exhibited and preserved. Liang's implementation of the flat cocoon technique as a semi-shamanic act in ritual practices is challenging to convey within current exhibition models. In contemporary art spaces, audiences typically encounter video recordings of performances such as Cloud Mirror or on-site installations, where silkworms – the main ritual practitioners – have already completed their work and left the art scenes. As a result, viewers are often excluded from the live processes of growth, healing, meditation, or contestation and remain insufficiently informed about the enactment of ritual actions. Liang's silkworm art thus calls for innovative methods of audience engagement, enabling viewers to experience the live ritual processes more fully and appreciate their ritual efficacy.

For projects spanning over two decades such as 'Time and Permanence', how to engage the audience in a more enchanting and efficient way into this prolonged 'ritual of life' beyond just viewing the 'installation' at a certain moment of its growth as an 'end product' in the gallery space? I have briefly addressed this issue in collaboration with a hybrid artist-archivist-technologist Eugenia Kim (Kim and Xu 2019). We explored how to use mixed reality technologies in virtualising Liang's silkworm art for both presenting and preserving purposes, which can be contextualised in sericulture practice as an intangible cultural heritage spreading across the world from China. The proposed model combines data curation, archival science and digital humanities strategies to capture the motions, the ephemeral and meditative Daoist ambience, erratic changes in materials and organisms over time in Liang's silkworm projects. As we (2019) notice, Liang's artworks, as well as some other contemporary bio-art projects, feature biological growth, non-human agency, and interspecies sociality that challenge conventional notions of object, life, and materiality. In art exhibition and preservation, art objects have traditionally served as thresholds or conveyors of narratives, concepts, and ideas, distinguished by their physical boundaries. Liang's projects, however, blur these borders, transforming art objects into processual, durational, ritual-like events with non-human collaborators, demanding new modes of interpretation and engagement. When visitors and critics remain absent from the growing processes of such works and lack familiarity with the core repertoire of ritual art, the most nuanced, reflective details and liminal aspects of Liang's practice are often overlooked, sometimes leading to misinterpretation.

Traces of Life and Death

Among the installations and performances examined above, 'Broken Landscape' (2008–2016) is exceptional in Liang's oeuvre, showcasing his most virtuosic techniques and grappling with the multifaceted liminalities of the motif '*ziran*'. While spinning silkworms in Liang's silk-wrapped installations typically left the stage after completing their tasks, 'Broken Landscape' advances this practice by retaining the use of the flat cocoon technique while preserving the physical presence and life traces of the silkworms within the ritualistic site (Figure 4.6).

The artwork features a long, thin rectangular piece of silk suspended from the gallery ceiling like a cascading waterfall. The lifespan of countless silkworms – from larvae to moths – is documented on this yellowish, variegated silk fabric, which bears their faeces, urine, and eggs in their original forms as black dots and pale brown stains (Figure 4.7). The rectangular silk resembles the bound scrolls of ancient Chinese paintings, traditionally depicting landscapes or botanical subjects. Yet, the colours and figures in Broken Landscape are far more abstract, constituting a 'landscape' of silkworm life cycles – eating, shedding, excreting, spinning, mating, and spawning. According to the artist's statement (Liang 2008), the work serves as a metaphor for the destruction of natural landscapes in China through human intervention, hence the title 'Broken Landscape'. The Chinese character for broken, '*can*' (残), is

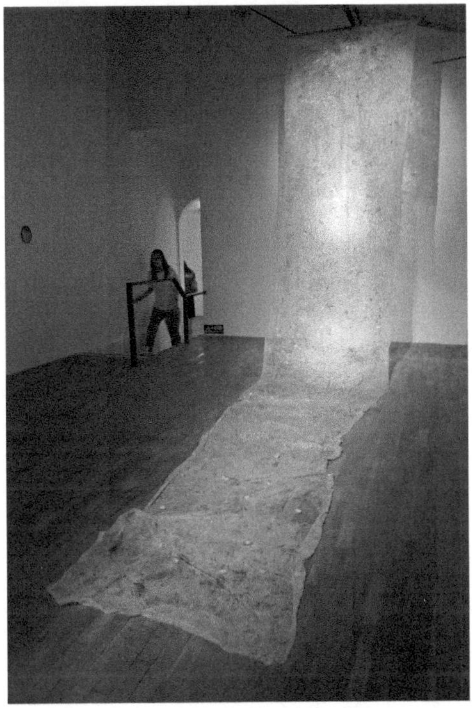

Figure 4.6 'Broken Landscape', 2016. Photo by author

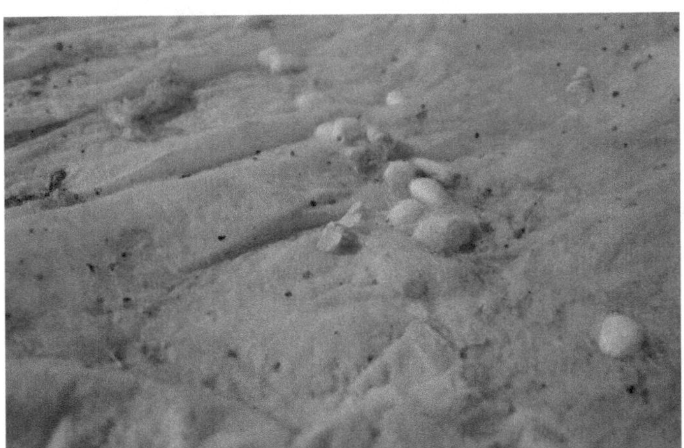

Figure 4.7 Life traces of silkworms on 'Broken Landscape', 2016. Photo by author

also a homophone for the word silkworm '*can*' (蚕), positioning the creature as the protagonist of the piece.

Silk, deeply associated with traditional Chinese painting, often serves as the canvas. In Broken Landscape, Liang moves away from uncanny materials such as glass, metal, or human skin, instead opting for a rectangular silk 'cloth' produced through the flat cocoon technique, similar to that used in making silk quilts. Unlike conventional reeled and woven silk, this fabric is thinner, lighter, airier, and more translucent. Silkworms were placed on this silk – spun by their deceased counterparts – to secrete additional flat silk filaments, thickening the material while leaving pale stains on its surface. This interplay between the painting and its canvas in 'Broken Landscape' recalls the artistic manoeuvres of 'Planar Tunnel' discussed earlier in the chapter, visually and conceptually. Yet, this work navigates even more delicate boundaries, transforming into a rite of *'ziran'*, where silkworms oscillate between being agentive performers and art objects. Silkworms serve as both the index of silk's existence – embodied in the rectangular fabric and newly spun layers – and as tangible entities spotlighting the generative, growing, and even vanishing nature of this material record of life.

The physical remnants of silkworm moths and clusters of silkworm eggs further highlight the tension between natural processes and human intervention in the creature's life cycle in 'Broken Landscape'. In the silk industry, vast numbers of silkworm pupae are boiled alive to extract silk. By contrast, in Liang's studio, silkworms – while required to spin flat cocoons in unnatural conditions – were allowed to complete their metamorphosis into moths. These flightless moths remained on the silk fabric until their natural death, embodying the culmination of their life cycle. Meanwhile, newly laid eggs scattered across the fabric did not hatch, entering a period of diapause – a suspended developmental state triggered by unfavourable conditions.

Diapause allows silkworms to synchronise their life cycles with seasonal changes, enabling survival during winters when mulberry trees cease to sprout. Over centuries of domestication, scientists have bred silkworm varieties with diapause periods ranging from months to a year. In sericulture, freshly laid eggs are often frozen to induce diapause and later 'awakened' in controlled environments. However, the eggs in Broken Landscape were left undisturbed at room temperature, remaining as tiny particles for months. In this way, the work embodies the Daoist principle of 'non-action' (*wuwei*) and non-interference, surpassing both sericultural practices and Liang's other flat cocoon art projects in its commitment to allowing biological processes to unfold and its subtle reverence for naturalness.

The Appropriation of Smell in Art Space

When the audience merely witnessed the 'final' appearance of this installation shown in art exhibition settings, those mutating and erratic processes of silkworms' physiology and morphology and the silk fabrics' tangible traits were seldom accessible to the audience and critics of 'Broken Landscape'. Though being incorporated into the field of art, silkworm's faeces in the form of round black pellets, for example, are regarded as a by-product in silkworm husbandry. 'Silkworm sands', as Chinese farmers euphemistically call them, contain pathogens and thus are supposed to be recycled by professional companies for fermentation and sterilisation before being processed into organic fertilisers. In sericultural practice, many thrifty farmers, nevertheless, ignore the experts' recycling instructions and shovel their silkworms' faeces into sacks and pour them directly into their own farmlands where mulberry trees or rice were planted. While the farmers considered the faeces much more nutritious than chemical fertilisers that should not be wasted, sericultural experts condemned such short-sightedness as silkworm sands would pollute the soil and water nearby and cause more silkworm diseases. Nor does Liang Shaoji intend to waste the silkworms' faeces, but his concern was not with their nutritional value or pathogenic danger, but with its visual form and symbolic potential as a heap of organic materials. As for the urine, silkworm farmers often sprinkled a thin layer of lime powder over the silkworm plaque or the ground of the silkworm room for the larvae, as a means of absorbing moisture and thus preventing disease. When a mature fifth-instar silkworm climbed onto a vertically hanging lattice cocoon frame and was ready to cocoon, it stuck its 'asshole' out beyond its measured three-dimensional cocooning space to drain the last drop of the faeces and urine without soiling the frame or other cocoons under construction by its fellow silkworms. During my fieldwork in Yizhou County in Guangxi, I once heard an elderly lady jokingly comment: 'silkworms are well civilised, and they don't defecate indiscriminately'. However, on the flat silk in Liang's 'Broken Landscape', the artist intentionally condoned and even invited such 'uncivilised' open defecation:

> Here I curated an environment for silkworms and a piece of silk cloth also made by silkworms to meet each other and stay together for a few days. The worms completed the artwork without me giving any order. I just waited along to see what would happen eventually and the realization of this artwork turned out to be an epiphany of what '*ziran*' is.

In 'Broken Landscape', the handling of olfactory factors was devious. The smell of silkworms' urine and faeces is hardly detectable to a human nose, according to my own experience living in the village raising silkworms. On the other hand, the odour of those droppings attracts silkworms' natural enemies such as bees and beetles as well as mosquitoes and flies which are commonly seen in rural farmhouses. Silkworm farmers in cultivating practice, therefore, have to do frequent cleaning and use screen doors and windows to keep those unwelcome insects out of the silkworms' rooms, while the artist didn't have to worry too much about this issue. The architectural design and management of museums and galleries are inherently inaccessible to most uninvited nonhuman creatures. As for the smell of 'Broken Landscape', there is another unnoticed dimension in the pre-purification process of this artwork. The most unbearable stink that would appear in the whole life cycle of a silkworm comes from the sick ones infected with a contagious virus commonly known as the 'pus disease'. It is a common disease of silkworms, and apparently, silkworms with pus disease had been deprived of access to 'Broken Landscape', an art scene free from smelly bodies and the white oozing pus.

I acquired this olfactory knowledge during my fieldwork in rural Yizhou as an apprentice to raise silkworms with local farmers. I often heard a saying by farmers and silkworm experts from the sericulture department of the local government: 'you walk by a silkworm house and the smell tells you whether the worms indoors are in good health or not'. Similar expressions would go like 'if silkworms are under good care, the room is aromatic' or 'good silkworms are aromatic'. I gradually built a valid connection between the signifier 'aromatic' and the linguistic signified smell when most silkworms in that village entered their fifth instar. I walked along the alleys, knocked into neighbouring farmhouses to have a look at their silkworms, and most importantly, sniffed hard with my nose. The idea of 'aromatic' naturally blew into my mind and took a root, when I had sensed the whole spectrum of the smell of silkworms, not only the 'aromatic' but also 'not too good' and 'terrible smelling' ones within a few days. I haven't encountered any similar smell like that during my entire life experience. When farmers use a too general adjective 'aromatic', or in Chinese '*xiang*' (香) to denote the very familiar scent in their daily life, it used to be impossible for me to comprehend such expression. This cultivated sense is only communicable in certain situations with people of certain experience. One needs to have experience in raising silkworms in sufficiently large amounts to pick out the aroma as an index of silkworms' health. Many elementary school students in China have had the experience of raising silkworms at home when their teacher of nature science distributed five to ten silkworms for each pupil to keep in a paper or tin box. This kind of small-scale cultivating experience would not help to provide an idea of what 'aromatic silkworm' smells like. Rather, working within a community of silkworm farmers might faster the acquisition of such olfactory knowledge by making comparisons between various silkworm rooms of each family.

Sense of smell is important in caring for silkworms. A 'bad smell' in a silkworm room is a simple index of something wrong with silkworms' overall condition and reminds the feeders to be cautious and act accordingly. The excrement of the silkworm, including faecal matter and urine, generally has no detectable smell to human noses. However, if this organic matter is fermented because the room is too humid, the odour indoors would be slightly unpleasant. Farmers should then open the windows and doors to ventilate or use other methods to adjust the microclimate within their farmhouse. When the room became extremely smelly, farmers were aware that some silkworms had been infected by a cytoplasmic polyhedrosis virus commonly called 'pus disease'. The infected silkworms would crawl about, bobbing their heads around. The oozing pus on its body would not only exude contagious pathogens but also emit a disgusting odour. A farmer described those pus-filled silkworms 'as if a poisoned person with a messed-up brain and fishy smell'. As the lifespan of a silkworm is far too short, with a total lifecycle ranging from 6 to 8 weeks, it would be too late for pills or injections to take effect after diagnosis. A silkworm with 'pus disease' needs to be picked up and discarded to forestall the spread of the disease among tens of thousands of its healthy peers. Olfactory sensation thus provides indispensable information for farmers to timely adjust and cope with adverse factors to silkworm's wellbeing.

Liang Shaoji almost certainly possesses this olfactory knowledge about silkworms though the viewers of 'Broken Landscape' are hardly likely to be aware of what had been eliminated from the ritual scene in the gallery. The silk textile in 'Broken Landscape' is visually messy and grubby, but the olfactory presentation of it is still pure, clean, and inoffensive, as sick stinking silkworms had been discarded and forbidden by the artists to enter into the arena of contemporary art. Waste and dirt have been materially appropriated, sensorily purified, and conceptually challenged by Liang's artistic strategies. It is notable that human being's fear of stains, death, and disease still implicitly hovers around like a phantom in art fields. The underlying filtering processes of 'Broken Landscape' guarantee the art space to be odourless, disease-free, clean, and elegant.

The Ultimate Rite of '*Ziran*'

When discussing 'Broken Landscape', Liang Shaoji explained his true visual source of inspiration for this piece of work. When he reared silkworms in Linhai village with local farmers, the yellow urine stains of silkworms suddenly reminded Liang of the surfaces of ancient Chinese paintings. Both paper and silk were abundantly applied as the ground material for Chinese painting through history, sometimes with glue treatment to improve the canvas' water resistance. After hundreds or thousands of years of oxidation, most existing Chinese paintings from ancient times take on a yellowish undertone. The artist's association between yellow fabric and Chinese painting has a cultural specificity. Likewise, viewers with some basic knowledge of Chinese painting would be more likely to recognise the visual resemblance between 'Broken Landscape' and typical Chinese landscape paintings and thus understand the connotation of its title. It wouldn't be difficult for those who have seen some ancient Chinese landscape paintings to appreciate the visual aesthetics of 'Broken Landscape', namely, its shape, colours, patterns, and textures.

However, when I finished my trip to Shanghai where this artwork was on display at an art exhibition 'Growing' at Chronus Art Centre in April 2019, I went back to Guangxi Province and continued to conduct my field research on sericultural practice in Yizhou County. It was serendipity that I stumbled across another version of 'Broken Landscape' in an unexpected place. Yizhou

Silkworm Breeding Station in the Hechi City of Guangxi is a government-funded silkworm egg production facility. Fresh cocoons were bought and carefully cut open; silkworm chrysalises undergoing a transformation from caterpillar to adult moth were then separated into two groups of male and female. Till the metamorphosis was completed, moths mated with each other of the same or a different variety according to the breeding requirements. The eggs laid by the female moths were processed and put into the market as a commodity. It was around 3:30 pm that afternoon, the time when the silkworm pupae started to spread their wings into moths according to this species' biological clock. Workers at the breeding station were busy laying a piece of yellow cotton cloth with round holes of five centimetres in diameter on each bamboo plaque, on which silkworm pupae were already densely placed. Adult moths who finished their morphological transformation gradually crawled up through the round holes onto the yellow-brown cloth, waiting to mate with another group of moths of the opposite sex (Figure 4.8).

A female worker explained the function of the cotton cloth with many round holes. The newly metamorphosed mature moth will urinate, and without the cotton cloth to absorb it the microclimate in every plaque would be too wet and challenging for other semi-pupae to spread their wings and become adult moths. The cotton clothes were washed, dried, and then reused repeatedly, and as a result, were stained dark yellow and brown by the urine of silkworms. I spent some time in the mating room of silkworms and squatted down to get a closer look at the cotton pieces on every plaque, with my camera zooming in on the texture of the cotton and the fluttering wings of

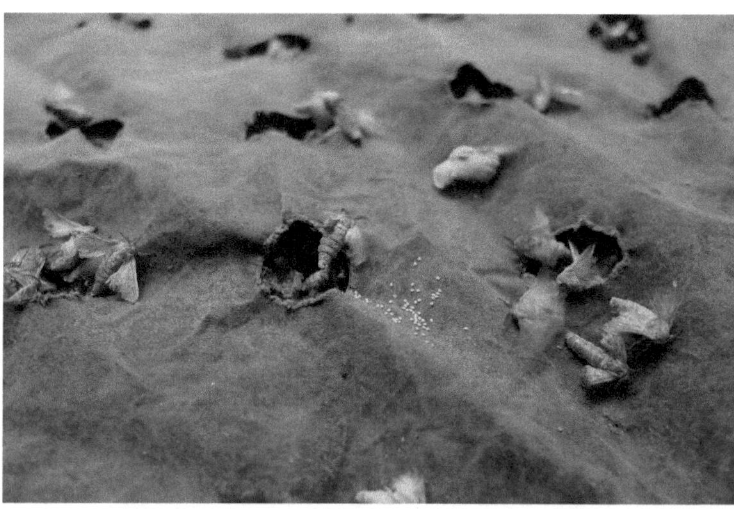

Figure 4.8 Cotton clothes used at Yizhou Silkworm Breeding Station. Photo by author

active moths. There were other cleaned cotton clothes drying under the sun on the second-floor balcony of the facility building. Workers at the station used them with no extra care, but I was thinking of another piece of fabric, also yellowish with the life traces of silkworms, which was placed in a gallery space at that moment as a durational rite of '*ziran*'. A pile of cotton clothes in the breeding station is as visually appealing as 'Broken Landscape', but they are still functioning day by day, closely involved in the final stage of silkworms' social life.

The symbolic and visual parallels between these two fabrics – silk and cotton – and their sharply contrasting roles in human society reflect the concept of 'frame' in social sciences. Frame theory, introduced by Goffman (1972), suggests that human actions are often shaped by the 'frames' that define their contexts and associated expectations. For instance, social roles prescribe specific behaviours for individuals based on institutionalised norms. Crucially, these frames often remain invisible unless they malfunction or deviate from expectations. In the art world, the frame plays an essential role in elevating objects to the status of artworks. It can take the form of a physical frame, a glass case, a gallery space, or even the institutional recognition of an artefact as 'art'. This framing mechanism helps explain why Liang Shaoji's silk piece in 'Broken Landscape', created in collaboration with silkworms, evokes far more attention and emotional response than the utilitarian cotton cloth used in silkworm egg production. 'Broken Landscape' being appreciated as an artwork elicits its hermeneutic potential, while the cotton cloth, devoid of such framing, remains functional and unnoticed.

However, interpreting 'Broken Landscape' as a semi-ritual art rather than merely a conceptually and visually innovative 'landscape scroll' reveals an even starker distinction between this silky artwork and the utilitarian cotton fabric. From this perspective, 'Broken Landscape' functions as a ritual of '*ziran*', where the boundaries between art object and performer dissolve. Silkworms take centre stage, their life cycle inscribed directly onto the fabric. Under pre-set conditions designed by the artist, they shape the silk's material qualities while simultaneously acting as shamanic agents in the ritual process. The silk itself assumes dual roles: as a physical trace of the silkworms' ephemeral existence and as a medium carrying the artist's intent, the ritual's temporality, and its own sensory and material properties.

The silk's agency is most profound when seen as an extension of the silkworms' bodies, blurring the line between organism and object. It serves both as evidence of silkworms' fleeting physical presence and as a vessel for their transcendence. Though the silkworms perish, their silk endures, extending their presence into a lasting artefact. This duality transforms silk from an inferential material into an active medium – one that not only transmits the artist's vision but also preserves the liminal states between life and object, intervention and spontaneity, dirt and purity, wildness and domestication, and art and artefact. 'Broken Landscape', therefore, transcends conventional categorisations like 'installation' in the contemporary art world. Instead, it should be understood as ritual art, rich in transitional states that challenge and expand the boundaries of art, materiality, and life itself.

References

Bray, F., 1997. *Technology and Gender: Fabrics of Power in Late Imperial China*. Berkeley: University of California Press.

Chen, G., 2013. Males' Leaf-picking in 'Pictures of Tilling and Weaving'. *Chinese Literature and History*, 09, 81–84.

Cheng, M., 2007. Animalworks in China. *TDR: The Drama Review*, 51 (1), 63–91.

Douny, L., 2017. Connecting Worlds through Silk: The Cosmological Significance of Sheen in West African Talismanic Magic. *Magic, Ritual, and Witchcraft*, 12 (2), 186–209.

Fan, Y., 2020. 河南巩义双槐树'河洛古国'遗址浅论 (On the Ruins of 'Heluo Ancient Country' of Shuanghuaishu in Gongyi Henan Province). 中原文化研究 *(The Central Plains Culture Research)*, 4, 15–20.

Fung, A., 2013. Condensation of Ritual Symbolism and Visual Culture: From Chinese *Liqi* to Contemporary Art Expressions. *In*: S. Leong and B.W. Leung, eds. *Creative Arts in Education and Culture*. New York: Springer.

Gladston, P., 2011. A Conversation with Liang Shaoji. *In*: *Contemporary Art in Shanghai: Conversations with Seven Chinese Artists*. Hong Kong: Blue Kingfisher.

Goffman, E., 1972. *Relations in Public*. Harmondsworth: Penguin.

Hafter, D., 2020. The Silken Tug-of-War in Eighteenth Century Lyons: The Gendered Nature of Knowledge in the Grande Fabrique. *In*: D. Schäfer, G. Riello, and L. Molà, eds. *SeriTechnics: Historical Silk Technologies*. Berlin: Max Planck Research Library for the History and Development of Knowledge, 33–42.

Hammers, R.L., 2011. *Pictures of Tilling and Weaving: Art, Labor, and Technology in Song and Yuan China*. Hong Kong: Hong Kong University Press.

Hansen, V., 2015. *The Silk Road: A New History*. Oxford: Oxford University Press.

Haraway, D.J., 2008. *When Species Meet*. Minneapolis: University of Minnesota Press.

Hiruma, K. and Kaneko, Y., 2013. Hormonal Regulation of Insect Metamorphosis with Special Reference to Juvenile Hormone Biosynthesis. *Current Topics in Developmental Biology*, 103, 73–100.

Kim, E. and Xu, F., 2019. Bringing the Tangible into the Virtual: Preserving Human-Silkworm Collaboration. *In*: *EVA London 2019: Electronic Visualisation & the Arts. Proceedings of a conference held in London 8th-11th July*. Presented at the EVA London 2019, BCS, The Chartered Institute for IT, 256–263.

Liang, S., 2002. About the Nature Series: Notes on Creative Work. *In*: W. Ai, ed. *Chinese Artists, Texts and Interviews: Chinese Contemporary Art Awards*. Hong Kong: Timezone 8, 41–46.

Liang, S., 2008. Notes on the Creation of 'Broken Landscape' [online]. *ShanghART Gallery*. Available from: https://www.shanghartgallery.com/galleryarchive/texts/id/10010.

Lu, X., 2001. 地母之歌: 中国彩陶与岩画的生死母题. 上海: 上海文化出版社.

Onaga, L.A., 2013. Bombyx and Bugs in Meiji Japan: Toward a Multispecies History? *Scholar & Feminist Online*, 11 (3). Available from: http://sfonline.barnard.edu/life-unltd-feminism-bioscience-race/bombyx-and-bugs-in-meiji-japan-toward-a-multispecies-history/

Poo, M., 1998. *In Search of Personal Welfare: A View of Ancient Chinese Religion*. Albany: SUNY Press.

Rong, X., 2003. 唐代宗教信仰与社会. 上海: 上海辞书出版社.

Stockard, J.E., 1989. *Daughters of the Canton Delta: Marriage Patterns and Economic Strategies in South China, 1860-1930*. Stanford: Stanford University Press.

Stockard, J.E., 2002. On Women's Work in Silk Reeling: Gender, Labor, and Technology in the Historical Silk Industries of Connecticut and South China. *In*: *Silk Roads, Other Roads: Textile Society of America 8th Biennial Symposium Proceedings*. September 26–28, 2002, Smith College, Northampton, Massachusetts, 272–281.

Turner, V., 1982. *From Ritual to Theatre: The Human Seriousness of Play*. New York City: Performing Arts Journal Publications.

Vallard, A., 2013. *Des humains et des matériaux: Ethnographie d'une filière textile artisanale au Laos*. Paris: Editions Pétra.

von Drathen, D., 2011. Liang Shaoji: Of the Immaterial and Its Corporeality. *In*: M. Bloemheuvel and T. van Kooten, eds. *Windflower: Perceptions of Nature*. Rotterdam: NAi Publishers.

Wang, Y., Zhu, X., He, K., Xia, J., and Meng, Q., 2016. A Review and Prospect on Breeding and Industrialization Application of Male Silkworm Varieties in Past 20 Years. *Science of Sericulture*, 42 (2), 189–195.

Xia, K., 2015. The Silk Shadow of Chora-topia [online]. *ShanghART Gallery*. Available from: https://www.shanghartgallery.com/galleryarchive/texts/id/8092.

Xia, K., 2017. Kiefer and Liang Shaoji: Differences of Modernity between East and West in Integrating Three Worlds. *Northern Art*, 12, 60–61.

Yin, Q., 2007. 丝绸之路与西域经济. 北京: 中华书局.

Young, S.H., 2013. For a Compassionate Killing: Chinese Buddhism, Sericulture, and the Silkworm God Asvaghosa. *Journal of Chinese Religion*, 41 (1), 25–58.

Young, S.H., 2017. "Bald-headed Destroyers of Living Things": Buddhist Identity in the Silk Cultures of Medieval China. *Asia Major*, 30 (2), 27–70.

Zhang, S. and Gao, H., 1999. 荥阳青台遗址出土丝麻织品观察与研究. 中原文物, 3, 10–16.

Zhao, F., 2005. *A Comprehensive History of Silk* (中国丝绸通史). Suzhou: Suzhou University Press.

Zhao, F., 2012. *Chinese Silk and the Silk Road* (錦程: 中國絲綢與絲綢之路). Hong Kong: City University of Hong Kong Press.

Zulueta, C.C., 2016. Two Artists, Two Ecologies, and a Shared Empathy Towards Non-Human Animals' Agencies: Yanagi Yukinori and Liang Shaoji. *Journal of Contemporary Chinese Art*, 3 (3), 377–387.

5 Anthropomorphic Kinship in Practice

In 1994, Liang Shaoji completed the material genesis of his artwork '*Baobao* (Babies)/Nature Series No.15', a meticulously crafted piece in which abundant silkworm cocoons were tenderly wrapped in vibrant red silk swaddling. It evokes the most intimate of human gestures: the cradling of newborn life. These silk-enshrouded white cocoons were destined to be scattered with deliberate spontaneity across gallery floors and public spaces, transforming mundane environments into sites of profound contemplation (Figure 5.1). Liang's later curatorial approach to 'Babies' was inherently contextual,

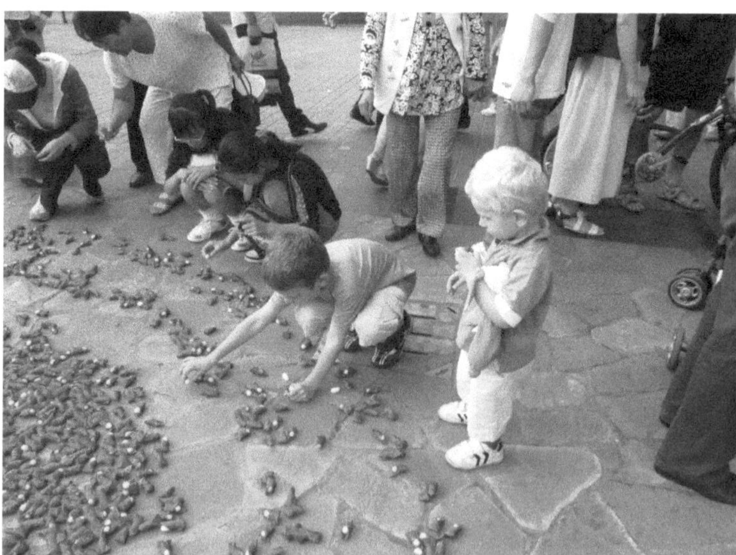

Figure 5.1 'Babies/Nature Series No. 15', 2009. Copyright: Liang Shaoji and ShanghART Gallery

DOI: 10.4324/9781003569091-5

each installation responsive to the social and geographical landscape of its presentation.

Following the devastating Wenchuan earthquake of 2008, for instance, Liang staged a poignant iteration in a public square near a shopping mall in Shanghai during the National Day holiday in October, the work resonating with layers of collective mourning and resilience (Figure 5.2). Years earlier, in 2002, the piece travelled to the 2nd Pančevo Biennial in Serbia – a region then emerging from the wounds of conflict, economically fragile yet yearning for cultural exchange. Liang, who rarely sells his silkworm art, gifted 'Babies' to the biennial's organisers in a symbolic act of empathy. Reflecting on the artwork's remarkable journeys, Liang recounted its transnational story:

> I invited the passers-by to come up and wrap the cocoons themselves. They could also adopt a silkworm baby, but I asked them to give the baby a name

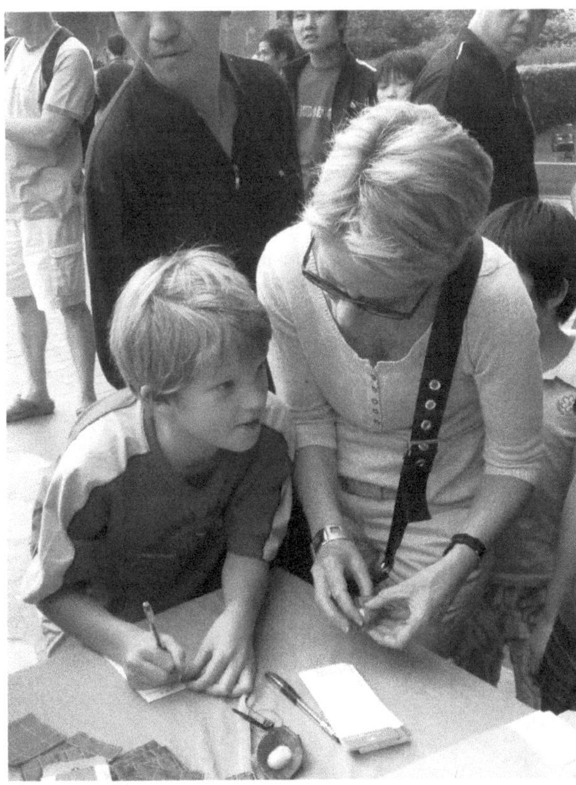

Figure 5.2 Audiences filling in adoption cards in 'Babies', 2009. Copyright: Liang Shaoji and ShanghART Gallery

> and write a blessing message on the adoption card. By the end of the day, the whole public space was plastered with cards. Even '*lao-wai*' (foreigners) joined in, some deeply moved, even donating money to these silkworm babies. '*Bao-bao*' resonated widely – in Canada (1998), Serbia (2002), China (2008) and the Netherlands (2009). Some viewers were even brought to tears, writing down heartfelt messages and wishes on the adoption cards.

The artwork 'Babies' employs an accessible visual language that transcends cultural boundaries, fostering universal comprehension through emotional and affective resonance. The silkworm's cocoon, a protective shield against pathogens and natural predators like flies and bees, ensures the pupae's safety during metamorphosis. In Liang's work, the red silk swaddling mirrors this protective function, adding a layer of shelter without burdening the delicate creatures. This visual metaphor is strikingly direct – audiences can intuitively associate the silkworm with a human infant, enabling the performative adoption ritual to unfold seamlessly across diverse cultural contexts.

Among Liang's body of work, Babies stands out. Unlike most of his 'flat cocoon' projects (see Chapter 4), where audiences encounter only the physical remnants of silkworm performances, this piece actively engages both human and non-human participants in real time. While most of his installations showcase silk flakes and wrapped objects left by silkworms after their brief life cycle, 'Babies' can be easily classified as 'socially engaged art' within contemporary art criticism but viewing it as a live adoption ritual offers fresh insights. Here, human spectators and silkworm pupae become active ritual participants in a shared performative act.

Adoption certificates preserved in ShanghART Gallery's archive reveal heartfelt 'Loving Words' from international participants – 'Fulfil dreams', 'I will try to [teach] her the Chinese philosophic Taoism', 'Long loving life', 'Always be well', 'Let's visit China together when you're old', and 'Laughter and happiness', among others (Figure 5.3). Many adopters may not have realised that the lifespan of a domestic silkworm rarely exceeds sixty days. Yet, this temporary bond, forged through the ritualised adoption, bridges cultural and biological divides. Building on this global and emotive reception, the silkworm-baby analogy invites deeper scrutiny into human-silkworm relations, in both speculative and pragmatic contexts. This chapter will unravel the often-invisible division of labour in silkworm cultivation, the human impulse to anthropomorphise silkworm care, and the moral implications of equating silkworms with human infants in both artistic and agrarian settings.

'Silkworm Baby' in Chinese Culture

For those familiar with Chinese culture, the artwork's title '*Bao-bao*' immediately evokes the phrase '*can bao-bao*' (蚕宝宝), which literally translates to 'silkworm baby'. In Chinese, '*bao*' means gem or treasure, and its reduplicated

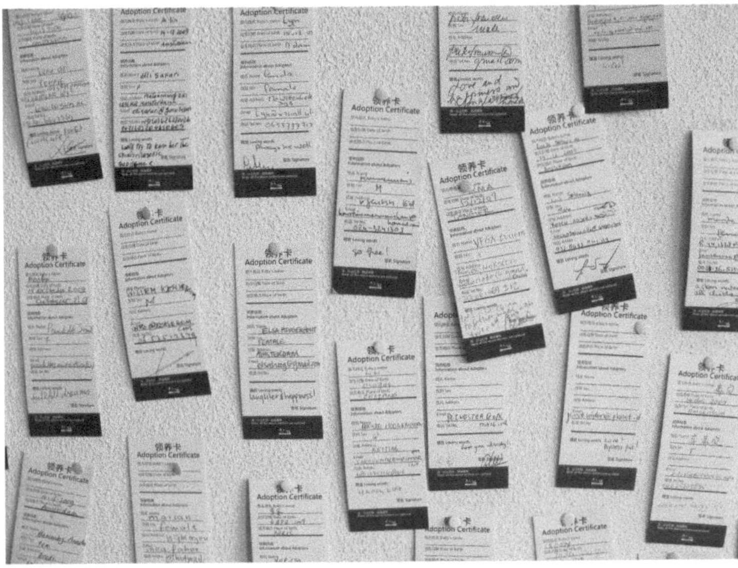

Figure 5.3 Loving words on adoption certificates in 'Babies', 2009. Copyright: Liang Shaoji and ShanghART Gallery

form '*bao-bao*' is an affectionate term primarily used for children, closely resembling the English word 'baby'. The term's usage extends far beyond infants. *Bao-bao* can be applied warmly to beloved individuals of any age – whether addressing a partner, child, or even a cherished pet. In Chinese culture, this endearing term reflects a deep sense of affection and value. A prime example is the beloved panda, often fondly referred to as 'panda *bao-bao*'. However, the most culturally significant pairing remains '*can bao-bao*', highlighting the silkworm's profound importance in Chinese cultural traditions.

One afternoon in spring 2019, I sat in an outdoor ground inside the Silkworm Breeding Station of Yizhou County, helping a female worker, Ms Liu, to cut cocoons. The silkworms had already pupated in the cocoons delivered to this breeding facility, and the workers' job was to cut off a corner of each one with a sharp knife, pour out the pupae and divide them into male and female groups for convenience in subsequent mating sessions. I examined the pupae on the palm of my hand and asked Aunty Liu sitting beside: 'Why are some pupae moving while others seem still?' She answered: 'Just like human babies, some crying and some laughing'. In her casual analogy, these silkworm pupae were still lovingly regarded as children, young and sometimes inscrutable, even though they had entered into the later stages of a silkworm's entire life cycle. In the summer of the same year, when I interviewed biologist Zhang at Guangxi Sericulture Research Centre in Nanning, he recalled the scene when he was giving guest lectures on silkworms to kids at the kindergarten his son attended. 'The children in that class knew me well, and they each held several silkworm larvae in hand to observe. One little girl came to me and asked: "Uncle Zhang, the silkworm *bao-bao* has so many feet. Does it wear shoes then?"' The biologist confessed that he was embarrassed as he couldn't think of a proper answer to these baby kids. 'This question about shoes always lingers in my mind', Zhang said.

The visual resemblance between silkworms and human children becomes most striking during their fourth and fifth instar stages. At this point, the larvae appear small, white, and chubby, with smooth skin reminiscent of an infant's delicate fingers. Biologist Zhang playfully underscored this resemblance by comparing artificial larvae feed to baby formula. Beyond mere physical appearance, these creatures demand care in ways startlingly similar to human infants. Silkworm larvae are extraordinarily vulnerable and require meticulous attention to nutrition, shelter, medical care, and protection from environmental threats such as pathogens and natural enemies. This profound dependency resonates deeply with feminist care ethics. Scholars argue that dependency is fundamentally a moral relationship, where all beings, human and non-human, require care to survive and develop (e.g. Donovan 2006; Held 2006). The term '*can bao-bao*' thus captures a profound interspecies kinship, suggesting that both vulnerable silkworms and human infants deserve moral consideration, love, and protection from those capable of providing care. In Guangxi, the world's largest cocoon-producing region, this kinship transcends abstract theory. Most farmers here share their living spaces directly with silkworms, creating an intimate domestic ecosystem. As biologist Zhang observed, silkworms become 'very important parts of our lives'. For many rural households, these creatures are more than just agricultural products – they are survival companions. The cocoon trade represents the primary, often sole, source of income, making the silkworm-human relationship a deeply reciprocal economic and emotional partnership.

This intricate bond is deeply rooted in Chinese cultural imagination. Literary traditions have long celebrated the silkworm's significance. Ancient origin stories, like that of Lei-zu – the Goddess of Silk – mythologise the creature's initial contact with human beings (Yuan 1985: 423). According to legend, a cocoon spun by a wild silkworm fell into a cup of hot tea belonging to Lei-zu, the main consort of the Yellow Emperor. Noticing the silk filament unwinding in the water, Lei-zu later pioneered the utilisation of cocoons to make fabrics. Over time, poets have immortalised silkworms as symbols of labour and sacrifice. Bai Juyi's Tang Dynasty poem poignantly captures this sentiment perfectly: 'Spring silkworms secrete silk until death' (春蚕到死丝方尽), portraying these creatures as embodiments of selfless dedication. Northern Song poet Zhang Yu critiqued class disparities through the silkworm's lens, writing, 'Those dressing in all delicate silk are not the ones who toil to raise silkworms' (遍身罗绮者/不是养蚕人). Mao Dun's political novella 'Spring Silkworms' (1932) documented the harsh realities of silkworm farming in Zhejiang Province during early twentieth-century social upheaval. More recently, silkworms have appeared as potent symbols in Gu Shi's sci-fi 'Poems and Distant Lands' (Gu 2020) and Wu Ming-yi's ecological novel 'The Man with the Compound Eyes' (Wu 2013). The contemporary art world has also engaged with the rich human-silkworm symbiosis in China. In Liang's 'Babies', the adoption ritual becomes a profound gesture, inviting participants to form a symbolic kinship with the swaddled silkworms and reflect on the ambiguous naturalness – or '*ziran*' – underpinning this species' domestication (see Chapter 2). However, once silkworm larvae encase themselves in ellipsoidal cocoons to undergo metamorphosis, they no longer require the everyday care and attention of their human raisers. Thus, while the adopters in 'Babies' committed emotional labour to the silkworms, they were spared the physically demanding and time-consuming cultivation tasks already carried out by silkworm farmers and the artist.

Forms of Labour behind Adoption

The juxtaposition between symbolic adoption and the practice of silkworm husbandry indeed creates a sense of dislocation: while the former fosters emotional labour in an artistic context, the latter demands sustained physical labour in agricultural settings. Liang navigates these two realms tactfully: his intensive care for silkworms aligns with his artistic objectives, yet his artworks remain deeply rooted in sericultural farming (see Chapter 2). Interestingly, even within the contemporary art world, Liang has not escaped the enduring cultural stereotypes tied to silkworm cultivation. The phrase 'Men farming and women weaving' (男耕女织), a deeply ingrained expression of the gendered division of labour in pre-modern agrarian Chinese societies, positions women as caretakers of silkworms and weavers of fabric. During my 2019 interview with Liang, he recalled the burgeoning stage of his career as a 'silkworm artist'. One comment that stood out to him came from a reputable Chinese art critic, who, in an informal setting, remarked, 'Ladies rear silkworms. You, a man, why bother doing this job?' Liang responded with laughter, seemingly dismissing the remark as harmless. However, the comment underscores the enduring cultural imprint of gendered labour divisions in Chinese sericulture, revealing how traditional norms persistently frame and challenge contemporary practices – even within the progressive realm of art.

Historical records and visual archives suggest that gender segregation in pre-modern sericulture was never as rigid as the phrase 'Men farming and women weaving' implies. While men typically handled tasks like leaf picking and moving tools, women performed the more meticulous care work of cleaning and feeding silkworms, akin to mothering (Bray 1997: 248, quoted in Xu 2021). In contemporary Chinese silkworm farming, men are also involved in tending

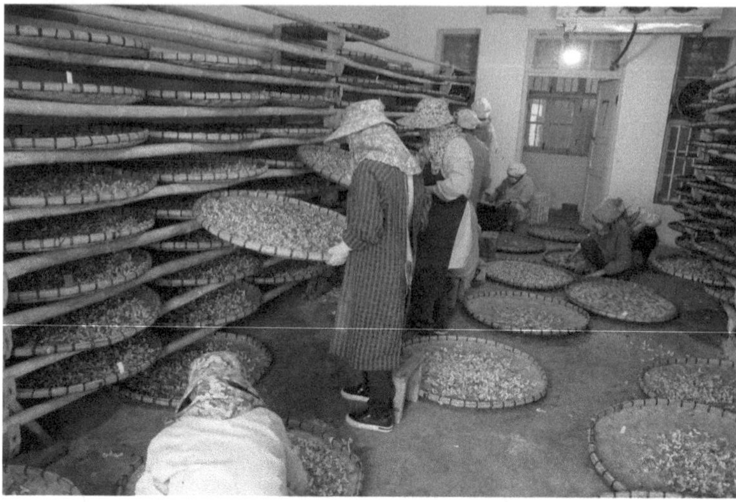

Figure 5.4 Female breeders helping with moth mating. Photo by author

mulberry gardens and feeding silkworms, though gender disparities persist in institutionalised settings. During my fieldwork in Yizhou County, Guangxi, I observed paradoxical patterns of gendered labour. In silk processing factories, sericulture departments, and silkworm egg production centres, managerial and research positions were predominantly held by men, while women undertook the bulk of administrative and operational work. At the Breeding Station, for instance, all silkworm breeders were middle-aged women, responsible for laborious tasks such as extracting pupae, assisting moths in mating, and handling freshly laid eggs. The working conditions were harsh: the mating rooms were filled with airborne scales from moths' wings, causing allergies and discomfort despite the use of masks and fans (Figure 5.4). When I asked why all the breeders were women, one worker laughed, replying, 'Which man would be willing to do this job?'

This dynamic reflects a subtler form of gender injustice, rooted less in the physical demands of labour than in entrenched power hierarchies. While women are excluded from agronomy education on the grounds that agricultural work is too strenuous, they continue to perform the majority of manual tasks in sericulture. Meanwhile, men trained in professional sericulture gain access to managerial and specialist roles, perpetuating a double burden of exclusion and exploitation for women in the industry. However, when Liang, as a male artist, outsourced tasks such as leaf-picking and silkworm feeding to villagers in Linhai, his approach reflected a division of labour shaped more by social roles than rigid gender norms. For Liang Shaoji, raising silkworms is undoubtedly an artistic endeavour, but for the farmers – both men and women –

the fate of the mature larvae, whether destined for the artist's studio or a silk-reeling factory, makes little difference to their daily cultivating work.

'*Yang*' as a Verb

Returning to 'Babies', the Chinese verb for 'adopt' – '*ling-yang*' (领养) – combines '*ling*' (领), meaning 'to claim or take', with '*yang*' (养), meaning 'to cultivate'. The concept of *yang* offers a nuanced perspective on cultivation that extends beyond mere sustenance. While it frames the act of raising silkworms, it also encompasses broader practices of nurturing and caregiving within Chinese thought. This linguistic and cultural lens ties the symbolic act of adoption in 'Babies' to its deeper roots in the philosophy of *yang*.

Unlike its English equivalents, '*yang*' is inherently purposeful, always implying a future-oriented approach to cultivation. In Chinese thought, *yang* is never a neutral act – it is always directed towards a specific outcome, whether practical or spiritual. This teleological essence of '*yang*' permeates diverse domains of life. Cultivating silkworms, children, plants, patients or even intangible qualities like Daoist '*qi*' (气) always entails an implied vision of the future. For instance, nurturing children often carries the expectation of future care in return (Liu 1998: 148), reverently serving one's parents fulfils filial duties (Knapp 2005: 114), and eating or exercising in a '*yang-sheng*' (养生) life-nourishing way seeks to maintain health and longevity (Kuo 2013). Yet this self-interest does not amount to pure selfishness, or egoism. Instead, it reflects a nuanced understanding of mutual benefit and interconnected relationships.

At its core, yang requires wisdom – a discerning ability to create conditions for growth and potential that benefit both nurturer and nurtured. In practices such as *yang-sheng* (cultivating health) or *yang-qi* (spiritual cultivation), the goal is not domination but a balanced, strategic care that nurtures possibility. In this light, the act of adopting a silkworm in 'Babies' transcends simple caregiving. It becomes a relational practice that invites a moral imagination of trans-species kinship and shared vulnerability. Whether raising silkworms, nurturing a child, or cultivating personal virtues, *yang* embodies a profound philosophical approach to life, one that sees cultivation as inherently strategic, hopeful, and deeply interconnected.

Cultivating Kinship with Silkworms

'Babies' is not the only project among Liang Shaoji's silkworm artworks where he explores the culturally conventional association between silkworm 'babies' and human offspring. Earlier in his career, Liang created 'Beds/Nature Series No. 10' (1993–1999), a work exhibited at the 48th Venice Biennale in 1999. In this piece, silkworms spun flat silk threads to cover charred engine coils shaped like miniature bed frames (Figure 5.5). However, in the

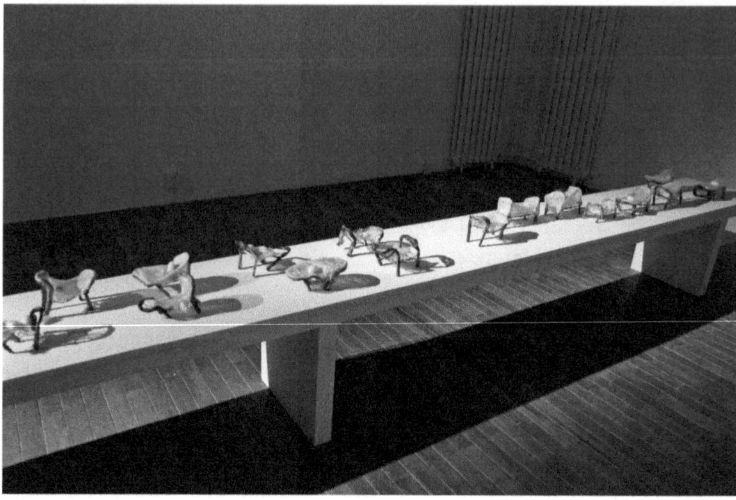

Figure 5.5 'Beds/Nature Series No. 10', 1993. Photo by author

art-making process, the silkworms were likely repelled by the smell of the metal but forced to secrete flat silk instead of forming ellipsoidal cocoons (see Chapter 3). In a conversation with me in his studio, Liang revealed that he initially attempted to twist the iron coils into the size of a human infant's cot. Yet he abandoned this idea, finding that the oversized frame undermined the sincerity of his care for the tiny silkworms. Reflecting on the final design, Liang explained, 'Only by making the bed frames to the scale of a silkworm's body size can I express the Daoist spirit of "*qi-wu*" (齐物)'.

The concept of *'qi-wu'*, derived from the Daoist text '*Zhuangzi'*, refers to equality among the myriad of things and beings. This philosophy resonates deeply in Liang's art practice, as art critic Yang (2018: 126) observes, where compassion and respect for insect life are central themes. Nevertheless, 'Beds' embodies a more anthropocentric, tenacious, and interventionist approach compared to 'Babies'. In 'Beds', silkworms were compelled to spin flat silk to wrap the iron bed frames, disrupting their natural instincts and depriving them of the opportunity to undergo metamorphosis within protective cocoons. Furthermore, the cot-shaped frames, though symbolically evocative, provided minimal shelter from pathogens or natural enemies and lacked the comfort that beds are typically designed to offer.

By contrast, 'Babies' employs a gentler, more empathetic approach towards silkworms. In this work, the silkworms were allowed to spin their natural cocoons while also being provided with additional, human-made shelters crafted from soft red silk rather than cold, hard metal. The swaddling clothes in 'Bao-bao' not only resemble raw-silk cocoons in form and function but also offer an extra layer of protection without imposing a burden on the silkworms. This thoughtful design highlights the artist's sensitivity and modesty. Where 'Beds' involved more unease, confrontation, and manipulation, '*Bao-bao*' reflects a more respectful and nurturing philosophy, better embodying the Daoist principle of '*qi-wu*'. The comparison between 'Beds' and 'Babies' underscores the paradoxically executed ethos of '*ziran*' in Liang's artistic strategies. Moreover, it reveals the sensitivity and ambiguity inherent in the analogical kinship between humans and silkworms, extending beyond the realm of art into the practical domain of sericulture where the intimate knowledge underpinning the daily care of silkworms reflects deeper lived anthropomorphic mindset.

Anthropomorphic Logic in Caring for the Silkworm

Over 3000 years of mutual domestication, breed selection, and adaptation have fostered a dependency where silkworms cannot survive without human care. In the Chinese context, this bond takes on a kinship-like quality, not only because silkworms are affectionately referred to as '*can bao-bao*', a term evoking the image and emotion of treasured and vulnerable human infants, as mentioned earlier, but also because of the specific cultivating strategies employed by human caretakers in their sericultural practice.

Silkworm babies have been carefully treated by farmers hosting and feeding them on a daily basis according to my observation when living and working with them in a village in Yizhou County, Guangxi. Experienced farmers all knew how to chop up the tender shoots from the tops of mulberry branches and feed them to the silkworm ants. When the larvae enter their fourth instar, matured enough to chew the leaves by themselves, the farmers will scatter whole leaves with their stalks to the worms (Figure 5.6). 'Just like babies, they should definitely eat finer food when they are small', as one villager commented. In addition to keeping the silkworms' rooms clean to prevent disease, it is best to maintain the indoor environment comfortable and well-ventilated with moderate temperature and humidity and to avoid too strong sunlight coming directly through the windows. Spring in Guangxi can be extremely wet for a few weeks and water can even bead up on the walls. The farmers once complained to me empathetically: 'It is too wet for us to stay at home and breathe just as the silkworms'. Silkworm larvae in their first and second instar are the most vulnerable and sensitive to the living conditions, and farmers, therefore, are advised to shower, wash their hands, and change the clothes they wear outdoors before feeding the worms. Some villagers were impatient with the fussiness of the junior silkworm babies, while some others showed understanding and tolerance: 'Well, we are also asked to wash the hands first before we held a new-born baby in the hospital, aren't we?' The anthropomorphic analogy between frail human infants and silkworm babies frequently worked in farmers' logic of feeding the silkworms.

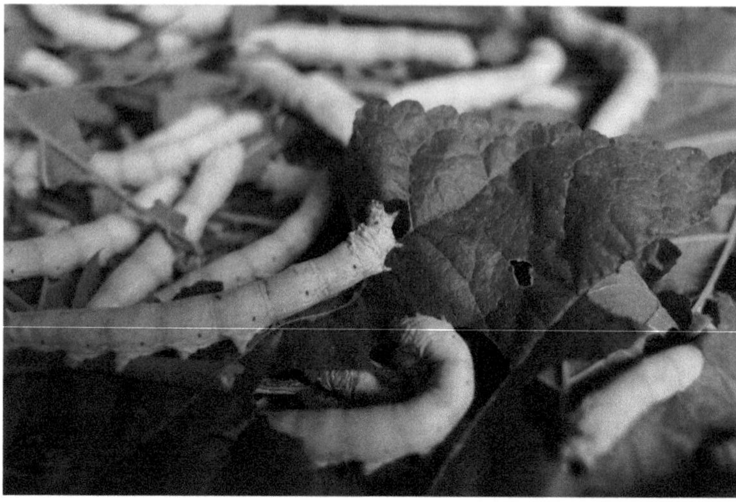

Figure 5.6 Silkworms eating mulberry leaves in the farmhouse. Photo by author

Such anthropomorphic logic was even more prominent in the doctoring practice of the silkworm in the agricultural context. When most villagers in Guangxi province dedicate a bedroom or half of the living room in their own farmhouses to raise domestic worms and become roommates for these creatures throughout their larval stage, a farmer should not only act as a host and a feeder but also must take on the responsibility as silkworms' 'family doctor'. Fragile and sensitive to its living environment, the silkworm behaves so fussily that it requires its human cohabitants to be diligent in cleaning and disinfecting the rooms and maintaining good hygiene habits. Otherwise, a wide range of pathogens, namely, viruses, fungi, parasites, pesticides, and even fluoride dust, can cause mass death of a whole batch of silkworms and render more than two weeks' caring endeavour of farmers futile. Disease management, therefore, has always been the keyword in both silkworm farmers' casual exchange of experiential strategies and also the research agenda of many bio-scientists. Core operational steps in silkworm disease management include thorough disinfection of the environment of silkworms' bedrooms and all the rearing tools, fertilisation of mulberry trees to ensure nutrient-rich foliage, microclimate regulation of indoor temperature and humidity, and proper isolation of silkworm excrement containing possible pathogens in order to avoid contamination of farmland and water resources. Disease prevention measures in sericulture demand more delicate skills than many other domestic animal husbandries, such as chickens and cows, partly due to the short lifespan of silkworms. A much smaller time scale makes it more impractical and infeasible to solve the disease problems by vaccines as used for pigs and antibiotics for chicken.

One headache for silkworm farmers is the proliferation of fake silkworm medicines in the sericultural industry. Some unscrupulous traders claimed that their medicine can cure all silkworm diseases and spraying the medicine on mulberry leaves and feeding the worms in advance can prevent the attack of any disease. Though such advertisements confuse preventive with curative effects in their blurred rhetoric, they cater to the cognitive inertia of some farmers to 'treat diseases'. During my months' stay at Village X, I realised that the analogous mentality of silkworm farmers in their veterinary processes traversed the species boundaries between humans and silkworms. Farmers who are willing to feed the silkworms with preventive or curative drugs should not be labelled as ignorant, scientifically illiterate laymen easily fooled by charlatans. But rather, behind their caring and doctoring practice, lies the pharmaceutical logic of treating silkworm babies as human offspring.

Local farmers were not unaware of the poor quality of veterinary medicines on the market. The setback of purchasing fake silkworm medicine, nevertheless, has not dampened many farmers' enthusiasm for feeding their silkworms with therapeutic drugs. They just gave up the animal versions and then turned to medicines for human use. A farmer, Mrs Qin, once explained her reasoning to me:

> The quality of medicine for human use is definitely better, as they must be strictly tested before being put into the market. There won't be too many fake human drugs. It would be a big deal if a person took fake medicine and then died from it. For veterinarians, medical malpractice is not necessarily a severe accident, is it? If an animal dies, let it be. That's why there are so many fake silkworm medicines on sale.

It was antibiotics for humans that have been selected by farmers and used on the silkworms for disease prevention or treatment purposes, including chloramphenicol, erythromycin, and roxithromycin. Experienced villagers shared the observation that if the mulberry leaves are too wet, either because of early morning dew or raindrops, the silkworms would become particularly susceptible to disease when eating too many soaked leaves. One farmer further explained the dosing logic when we were chatting in his farmyard:

> We were afraid that the worms would have diarrhoea with too much water in their bodies, so we dilute one small bottle of antibiotic injection for treating human diarrhoea and spray it on silkworms' daily meals as a precaution. And human's antibiotics aren't too expensive. The price used to be two yuan per box and now is five yuan.

In rural areas in Guangxi, taking cheap antibiotics for diarrhoea is an economic, convenient, and effective survival tip for many farmers living in the countryside. When the expected lifespan is drastically reduced, from an average of decades for humans to just over 40 days for silkworms, therapeutic drugs that act on human bodies become preventive drugs for silkworms. The form of 'misuse' of antibiotics has transformed from 'unnecessary dosing for unrelated symptoms' on humans towards 'dosing even without symptoms' on silkworms. The significantly reduced disease resistance of silkworms after consuming too many wet leaves has been empathetically interpreted by a few farmers as an increasing tendency to have diarrhoea. Their analogical pharmaceutical usage of antibiotics for silkworms thus became reasonable, despite the temporal discrepancy between therapeutic and preventive effects brought about by the gap between two species' polarised lifespans. Farmers in the same or neighbouring village often exchange their experiences in dosing the antibiotics during their casual conversations, such as whether chloramphenicol would be more useful than erythromycin, and what would be the appropriate ratio of water to the drug. Those active intra-community experiential tips, at least, demonstrate that pragmatically dosing antibiotics to silkworms as a precaution against diseases does not have noticeable negative effect on these creatures in a short span of time.

As for the effectiveness of broad-spectrum antibiotics in silkworm disease management, it would be a difficult issue to quantify and draw definitive conclusions. This is partly due to the fact that there are too many variables in silkworm rearing practice that it would be impossible to trace a single causal train in one's failure or success of a batch. Nonetheless, the logic behind using human-use antibiotics on the silkworm is certainly problematic in terms of pharmacology when pathogens of major silkworm diseases in Guangxi are viruses and fungi, which cannot be killed by antibiotics. The farmers in Yizhou fed human-use antibiotics to silkworms as they purposefully tried to avoid the risk of buying fake silkworm-use drugs in the less strictly regulated veterinary drug market. They diluted the antibiotics with water and sprayed them on freshly picked mulberry leaves, as one farmer Mrs Qin once reasoned: 'When we give medicine to a child, we always break the tablets in half and only feed one half of it to a baby. The same goes for silkworm larvae'. Farmers very considerately reckoned with the concentration of the medication, diluting the antibiotics with water because the silkworms are delicate babies just as human infants.

Such anthropomorphic logic in silkworm husbandry reflects a deep cultural and emotional resonance that goes beyond mere practical care. Farmers' treatment of silkworms as 'babies' – manifested in their meticulous feeding routines, diligent disease prevention methods, and unconventional use of human antibiotics – denotes a shared sense of vulnerability and dependency across species boundaries. The widespread reliance on human-use antibiotics, despite their questionable pharmacological effectiveness for silkworm diseases, exemplifies how this analogical cognition drives pragmatic decisions rooted in a deep-seated cultural ethos of cultivation. This ethos, shaped by longstanding traditions of interdependence between humans and silkworms in Chinese sericulture, frames silkworms as delicate lives requiring extraordinary protection and nurturing. Contextualising Liang Shaoji's 'Babies' within this framework provides a richer understanding of the affective and philosophical undertones of this art project. It evokes not just the literal care of silkworms but also the broader moral implications of intimacy, kinship, and responsibility towards the vulnerable, bridging the realms of artistic expression and agricultural practice in the undercurrent.

Multispecies Ethics Beyond Art

In the semi-artistic and semi-pragmatic scenarios discussed above, anthropomorphism emerges as a constructive mode of fostering empathy and understanding between humans and nonhumans. As the plant scientist Myers (2015: 44–49) argues, when employed thoughtfully, an anthropomorphic approach can serve as a powerful 'lure' for recognising and engaging with different 'modes of attention'. Rather than dismissing anthropomorphism as a simplistic or anthropocentric fallacy, it can facilitate a richer dialogue with nonhuman life forms. This is evident in Liang Shaoji's 'Babies', where the silky swaddling cloth becomes a medium of reference, forging profound visual, material, and psychological connections between infant silkworms and human beings. The explicit title 'Babies' further amplifies this effect, using an affectionate term to elicit the audience's empathy towards silkworm's fragility and reliance on human care. While non-Chinese-speaking audiences may miss the deeper cultural connotations of '*can bao-bao*', the universal resonance of the word 'baby' ensures that this ritual event of silkworm adoption transcends cultural boundaries. Liang's innovative curation not only invites his audience to cherish and empathise with delicate lives but also bridges the gap between human and nonhuman existence through a shared emotional register.

However, engaging with silkworms – or any nonhuman beings – through anthropomorphism presents challenges, particularly concerning temporal scale and individual perception. For silkworm farmers, the brevity of the insect's 50-day lifespan might feel counterintuitive when determining the timing and dosage of silkworm medicines. Similarly, audiences encountering 'Babies' may misinterpret the lifecycle stage of the silkworms inside the cocoons, perceiving them as infants rather than recognising the late stage of their metamorphic journey. These misalignments in perception raise deeper questions about how humans relate to individual animals versus animals as collectives. As Milton (2005: 260) observes, understanding others may derive not solely from humanness (anthropomorphism) but also from egomorphism, which prioritises personal experience as the basis for understanding, connection, and communication. Milton's distinction between 'human-centric' and 'self-centric' perspectives prompts further reflection on the analogy between silkworms and human babies. Do audiences participating in silkworm adoption performances view the insect as 'like me', 'like my child', or as 'like a human being'? Similarly, do silkworm farmers relate to the creatures they raise through a personal, self-referential lens, or do they engage with them as functional members of a collective species?

For artists, scientists, and silkworm farmers with long-term experience interacting with silkworms, the species is almost always perceived in its plural form – as collectives, as groups, as batches of hundreds or thousands of individuals. In these contexts, the relationship is inherently one-to-many, with knowledge and experience accumulated through repeated interactions, albeit not with the same individuals. By contrast, in the artistic setting of Liang Shaoji's 'Babies', the adoption ritual reframes this relationship as one-to-one: a single human being symbolically adopts a single silkworm baby. This approach mirrors more familiar and comfortable forms of interaction, communication, and kinship-building, even though these adopters are not required to assume the daily responsibilities of silkworm care, such as feeding or cleaning. Drawing on my own direct experience of cultivating silkworms, I adopt a distinct perspective on Liang's work, one that highlights the valuable dilemmas and paradoxes regarding anthropomorphism and its moral implications in fostering cross-species kinship.

Regarding anthropomorphism and moral imagination, an ancient Chinese example of human-silkworm relations offers a thought-provoking perspective. The sericultural industry, which involves the mass killing of silkworm pupae, stands in stark contrast to the Buddhist principle of non-violence, rendering it seemingly immoral. However, as Young's historical research (2013: 40) reveals, there was a significant shift in Buddhist attitudes towards sericulture during the Tang Dynasty. In the hagiography of Monk Zhiju, silkworms are no longer depicted as hapless victims of human greed for silk but as 'self-aware agents of compassion' and 'the bodies of bodhisattvas transformed'. These silkworms are portrayed as willingly and generously sacrificing their lives in a noble act to alleviate worldly suffering and destitution. In this reimagined narrative, silkworms are not killed by human hands; instead, they take the initiative to give their lives, akin to the Buddha's own acts of self-sacrifice, elevating their deaths from mere economic necessity to a soteriological and ethical offering (2013).

This historical account evokes deeper reflection on role of anthropomorphism in trans-species moral reasoning and justification. Beyond serving as a channel to elicit empathy, attention, and care, can anthropomorphism also operate as a narrative strategy – one capable of reframing complex moral and ethical dilemmas? Furthermore, does the ultimate purpose of such narratives incline towards altruism, or do they risk slipping into egoism? In this context, might the use of anthropomorphism in imagining cross-species moral communities be an empowering gift, fostering a more inclusive ethical vision? Or does it risk becoming a dangerous trap, reinforcing human-centred biases or, worse, descending into self-serving sophistry?

Engster in his ethical inquiry into the abolition of factory farming, declares that '[w]e do at least owe animals moral sympathy and take on moral duties to care for animals when we make them dependent upon us for their survival and basic wellbeing' (2006: 522). This moral obligation, grounded in a relationship of dependence, applies to most cases of domestic poultry and mammals. As Engster presents, a central dilemma in the factory farming debate is whether death is intrinsically morally inferior and whether an animal's life under attentive care until the moment of slaughter can justify its commodification (2006). The particularity of silkworm husbandry lies in its partial divergence from industrial farming practices. Unlike poultry or cattle, sericulture has not yet been scaled up into large-scale factory farms; instead, silkworms are typically raised in small-scale domestic farmhouses. Moreover, while meat constitutes an essential survival resource for many, silk is primarily a luxury material, and rejecting its production might seem a morally straightforward choice. Yet, domesticated silkworms cannot survive without human care, leaving us with a moral paradox: does our duty to care for silkworms absolve us of the ethical implications of their deaths for non-essential purposes?

Rather than engaging in ethical critiques of the silk industry, I focus on the moralities that emerge through everyday human-silkworm interactions across artistic, scientific, and agricultural contexts. Returning to the intimate adoption ritual in Liang Shaoji's 'Babies', it is striking how anthropomorphism, often dismissed as reductive or anthropocentric, can, in fact, open new pathways for ethical reflection and imagination. In human-silkworm sociality, the logics of care, attention, and empathy intertwine with the instrumental rationalities present not only in Liang's art but also in agricultural and industrial practices. Anthropomorphism, or egomorphism, invites us to envision multispecies coexistence beyond utilitarian frameworks, foregrounding empathy and shared vulnerability instead. Although such an analogical approach remains contentious in multispecies research, its potential to deepen moral inquiry is far from fully explored. In grappling with multispecies ethics, one must embrace the inherent uncertainties and ambiguities of cross-species interactions, recognising that human understanding of nonhumans may often rely on metaphor or even illusion. Perhaps the Daoist spirit of '*ziran*', one that loosely values fluidity, adaptability, and the acceptance of ambiguity, offers not only to ease and openness but also courage and a meaningful lens to navigate mundane encounters in ethically imaginative and practically compassionate ways.

References

Bray, F., 1997. *Technology and Gender: Fabrics of Power in Late Imperial China*. Berkeley: University of California Press.

Donovan, J., 2006. Feminism and the Treatment of Animals: From Care to Dialogue. *Signs*, 31 (2), 305–329.

Engster, D., 2006. Care Ethics and Animal Welfare. *Journal of Social Philosophy*, 37 (4), 521–536.

Gu, S., 2019. *Poems and Distant Lands (为了生命的诗和远方)* [online]. *Current Futures: A Sci-fi Ocean Anthology*. Available from: https://go.xprize.org/oceanstories/poems-and-distant-lands/.

Held, V., 2006. *The Ethics of Care: Personal, Political and Global*. Oxford: Oxford University Press.

Knapp, K.N., 2005. *Selfless Offspring: Filial Children and Social Order in Medieval China*. Honolulu: University of Hawai'i Press.

Kuo, C.P., 2013. Pigs, Pork, and Ham: The Practice of Pig-Farming and the Consumption of Pork in Ming-Qing China. Ph.D. Dissertation. New York University, New York.

Liu, Z., 1998. *Medical Sociology of Women (女性医疗社会学)*. Taipei: FemBooks.

Mao, D., 1932. 春蚕 (Spring Silkworms). 现代, 2 (1), 9–26..

Milton, K., 2005. Anthropomorphism or egomorphism? The perception of non-human persons by human ones. *In*: J. Knight, ed. *Animals in Person: Cultural Perspectives on Human-Animal Intimacies*. Oxford: Berg, 255–271.

Myers, N., 2015. Conversations on Plant Sensing: Notes from the Field. *NatureCulture*, 3, 35–66.

Wu, M., 2013. *The Man with the Compound Eyes*. London: Harvill Secker.

Xu, F., 2021. Gender Division of Labour: From Human to Silkworm in Sericultural Practice. *Proceedings of Politics of the Machines - Rogue Research 2021 (POM 2021)*, 220–225.

Yang, J., 2018. Dao of the Silkworm: The Ecological Thrust of Liang Shaoji's Nature Series. *New Arts: Journal of the National Academy of Art*, 39 (6), 119–127.

Young, S.H., 2013. For a Compassionate Killing: Chinese Buddhism, Sericulture, and the Silkworm God Asvaghosa. *Journal of Chinese Religion*, 41 (1), 25–58.

Yuan, K.,1985.中国神话传说词典.上海: 上海辞书出版社.

6 Epilogue

Working Towards '*Ziran*'

On a brisk morning in April 2019, I was about to take a ride to Linhai County, which lay an hour from Tiantai and offered the closest high-speed rail station for accessing the Tiantai area. Liang saw me off by the entrance of Tiantai Museum, asking about the progress of my doctoral research. I shared insights from my time in rural Guangxi, where I had lived among farmers, learning to cultivate silkworms. He was curious to know how many batches of silkworms Guangxi farmers could raise annually, and I explained it was about 11 – slightly higher than in Zhejiang Province, thanks to Guangxi's warmer climate providing a longer rearing season.

'Perhaps I could transplant my studio to Guangxi, then I'd have more time to work'. Liang joked. But almost immediately, he dismissed the idea with a self-deprecating laugh, citing the challenges of transporting delicate artworks from a remote region to the megalopolitan exhibition spaces, and more importantly, the physical demands that heavier workloads might strain his ageing body.

Yet, in my interactions with him, Liang Shaoji radiated an almost dazzling vitality – spiritual and physical energy that defied the typical markers of age such as weakness and decay. When speaking of his creative work, his voice always resonated with a robust clarity, his thoughts articulated with precise recollection of abundant details.

The night before, I spent three engaging hours in freely flowing conversation with Liang and his friend Xinhua, a ceramic artist who also lives in Tiantai, almost until midnight. At one point, Xinhua asked, 'Are there any ballads or verses associated with silkworm farming?'

Liang replied, 'I haven't really heard of any. In terms of superstitious practices, when silkworm eggs first begin to hatch, there's usually a ritual to honour the silkworm deity'.

'Ceramics also involve rituals. There are ceremonies when opening a kiln. I've even heard different versions of the old stories about live human sacrifices to kilns'.

After some thought, Liang added, 'It's because *ziran* isn't entirely comprehensible, so at times like these where outcomes are unpredictable, people

DOI: 10.4324/9781003569091-6

turn to mythological spirits. In the countryside, I've heard that rituals used to be common, but they're much less so now. For one, in coastal areas, the rise of alternative livelihoods means that silkworm farming is left to the elderly – young people no longer learn it. Ceramics are in a slightly better position; more youngsters study ceramics, but none seem interested in silkworm farming. And then there's the advent of synthetic fibres, which pose a significant challenge to natural silk'.

This exchange reminded me of a broader reality: the decline of sericulture across China. In rural Guangxi, as in Zhejiang, silkworm farming is predominantly the domain of middle-aged and elderly individuals. This underscores a broader challenge faced by the silkworm industry – its marginal position within the economic mainstream. Perhaps, in this regard, Liang Shaoji's art with silkworms shares a quiet affinity with sericulture itself, both existing in a solitary and peripheral space, distanced from the noisy currents of popular interest.

Karmic Rapport and Shared Aliveness

When reflecting on the circumstances that led him to remain in Tiantai for his artistic practice in the 1990s, rather than moving to major cultural hubs like Hangzhou or Beijing, Liang Shaoji framed his decision as being guided by what he called 'the will of life'. While shaped in part by conditions beyond his choosing, he embraced this path with resilience, this resilience forms the foundation of his intimate bond with silkworms, which has sustained his thirty-year-long artistic journey with these creatures.

> I can't quite articulate it… why I feel such a fateful affinity with silkworms. It's like the temperament of Tiantai County itself, and the character of silkworms – they resonate with me in a way that feels almost predestined. We just click. This sense of connection compelled me to stay here, to continue creating. In the end, what we call an artistic style, I think, naturally flows out of such a rapport. It isn't devised, it isn't constructed, and it certainly isn't imitated. My works carry a certain bitterness, my life holds its shared of loneliness, but also a sense of resilience. In a way, I feel my life has revealed itself through my experiences and my art.

The Chinese phrase '*tou-yuan*' (投缘) employed by Liang, resists precise linguistic transposition into English, yet intimates a literal notion of being 'thrown into fate' literally, thus conveying a spontaneous, natural, and almost karmic aura inherent in his rapport with silkworms. This expression has a mystical, personal, and poetic sentiment, and such personalness is also manifested in the Chinese title of Liang's large-scale retrospective exhibition in Power Station of Art in Shanghai in 2021, '我蚕 蚕我' (Me-silkworm, silkworm-me), while its English title 'A Silky Entanglement' failed to capture this nuanced undercurrent. Throughout three decades, the lives of nearly a million silkworms flow through Liang Shaoji's life like converging rivers. This long-term one-to-many relationship is semi-romantic, semi-religious, and intensely personal and therefore too unique to be carelessly slotted into any animal-involved art genres or abstract philosophical jargons. It is indeed a silkworm-Liang relationship that demands instead an immersive, phenomenological understanding that emerges from its own intrinsic logic. The resilient aliveness, that Liang perceives in silkworms – a karmic resonance that vibrates with the ineffable expression of '*tou-yuan*', crystalises the quintessential enchantment permeating both his artistic odyssey and his innermost being.

In the preceding chapters, I argued that art critics have underestimated the significance of Liang Shaoji's silkworm-cultivating processes in grasping the multi-layered ethical and philosophical implications of his art. Here, I propose a more radical idea: Liang's 'cultivating-in-making' relationship with silkworms constitutes art itself. What has been overlooked is not just the preparatory work preceding the final art objects or presentations, but the artistic co-working process and ongoing relationship as art – a relational, affective, and moral practice is imbued with ritualistic, experimental, and cosmological dimensions.

The distinctive character of Liang's artistic collaboration with silkworms lies in its radical departure from the more transient employment of nonhuman capacities in contemporary art akin to unpaid outsourcing. The sheer duration and intensity of Liang's focused engagement with a single species effectively propels his artistic practice beyond existing analytical frameworks of human-animal relations in contemporary art, such as new materialism and animal studies. Instead, Liang's silkworm art emerges as a deeply private ontological exploration that intrinsically resists top-down imposed posthuman cosmological schemes. While Liang's cosmological exploration is subtly influenced by Daoist thought, it moves beyond traditional interpretations of '*ziran*'. Instead, his silkworm art has spotted, whether intentionally or not, anti-structural paradoxes around the ecological motif '*ziran*', namely, between the modern and the preindustrial, organism and object, growing and making, art and artefact, intervention and adaptation, materiality and spirituality, artificiality and naturalness, anthropocentric manipulation and creative intervention, etc. Moreover, Liang's art illustrates how these liminal encounters compel humans – Liang included – to make emotional and moral efforts to think and to take action. In this sense, Liang and silkworms do not merely work within, on or as part of '*ziran*', but collaboratively work towards it, enriching both its implosive complexities (cf. Dumit 2014) and its ethical potentials.

Revisiting Ritual and Liminality

This book provisionally interprets Liang Shaoji's silkworm projects as ritual art, drawing upon a rich body of scholarship on ritual studies. Like art, rituals can have transformative consequences and practical efficacy beyond their presentational and representational functions (cf. Morphy 2013: 655). Most anthropological studies on ritual – particularly its efficacy – focus on its macro-level functions within a community (e.g. Geertz 1973; Leach 1973; Rappaport 1968). These range from modelling social values to promote cohesion in initiation rites (Durkheim 1912) to safely releasing social tensions through 'rites of rebellion' (Gluckman 1965). Moving beyond approaches that seek statistical models or symbolic structures to infer the influence of rituals (e.g. Rappaport 1968; Geertz 1973), this book primarily adopts Alfred Gell's (1998) framework, which foregrounds the agency of both the initiator (artist) and the recipient (audience) on a more interactional and cognitive scale. In Gell's approach, ritual art functions by reorienting people's mental states and thought processes. Rather than treating art objects within rituals as conveyors of symbols, Gell highlights the materiality of both humans and nonhumans. Their physical traits and capacities influence each other through the mediation of artistic techniques. Liang Shaoji's silkworm art is analysed within this framework, as his projects critically reflect structured social tensions and contemporary quotidian life (e.g. Gladston 2011; von Drathen 2011; Xia 2015; Zulueta 2016). His art 'plays creatively' on these tensions, much like rituals, to dissolve conventions (cf. Ingold 2013: 342).

This creative and processual force of rituals, which can potentially enact social change, is a central element of the concept of 'liminality'. As a theoretical construct, liminality captures the critical and creative capacities of rituals (van Gennep 1909; Turner 1969, 1974). The term 'liminality' denotes the in-between spaces where changes and transformations occur. Liminality has been widely applied in anthropological studies of rituals to explore ambiguous and malleable situations. In his study of 'rites of passage', van Gennep (1909) emphasises ritual's capacity to alter social status. He identifies three phases of rites of passage: separation, transition, and incorporation. The transitional stage, or liminal zone, is a space of ambiguous and transformative socio-cultural identity and ontology. Building on van Gennep's work, Turner (1969) highlights the creative power of the liminal stage in rituals. Liminality, he argues, can subvert or negate established social categories, conventions, and structures. Rituals can generate insight, resistance, and potential social transformation by linking the bodily experiences of ritual practitioners with abstract thought. Turner (1974) later extends the concept of liminality beyond ritual scenes to socio-cultural life. In this expanded framework, liminality represents an anti-structural, continuous state that resists social hierarchies and structural divisions. Art, in this sense, can occupy a 'liminal' space in contemporary society by challenging categories, questioning traditions, and provoking reflection and creativity. While Turner's emphasis on the autonomy, creativity, and anti-structural potential of liminality has been criticised for romanticising change and portraying everyday life as overly routinised or static (Rapport and Overing 2000: 235), the concept remains an invaluable tool for interpreting practices that navigate the interstitial zones between tradition and innovation.

This theoretical foundation provides a lens through which Liang's silkworm art can be seen as bearing revolutionary potential by destabilising and reimagining the deeply rooted cultural-religious concept of '*ziran*'. His invocation of the multivocal '*ziran*' – encompassing notions of 'nature', 'naturalness', 'what is spontaneously so', and 'the way it intrinsically goes' – moves beyond a mere adherence to Daoist principles. Instead, his ritualistic engagement with silkworms and his life-long creative practices constitute a revolutionary probe into the fluid, destabilised meanings of Daoism-flavoured '*ziran*'. Liang's ritual art thus even occupies liminal spaces for critical reflection and perceptual transformation, with potential emancipatory implications for art history, socio-cultural practices, and ecological thought.

Ethnographic Research into Multispecies Labour

Throughout this book, I juxtapose silkworms raised in diverse practices – spanning contemporary art, scientific study, and agriculture – alongside the silk they spun, used for varied purposes. This approach seeks to foreground silkworms as active agents with roles that resist reductive categorisation when exploring the moral undertone of Liang's art. It also aims to move away from the dominant frameworks of STS studies and posthuman theories in studies of animal-involved art, which tend to explain away animals for understanding interpersonal dynamics, societal structures, or artistic labour. In these homogeneous camps, animals, plants, or microbes once again serve as 'mirrors or windows' for reflection on humanness despite those novel terminologies (cf. Mullin 1999). La Berge (2019), for example, insightfully draws on Marxist and neoliberal theories of labour to interrogate the power dynamics between human artists and their animal collaborators in socially engaged art, where the formers' 'overproduced, underemployed, and indebted' status gives more room for the latter's agency as cultural workers, artists and creators (2019: 122). Their primary argumentative goal, however, remains tethered to human-centric concerns – revealing how de-skilled and re-skilled animals in art illuminate the devaluation of human artistic labour, unwaged and decommodified, rather than questioning the status of animals as artists in their own right. Liang Shaoji's art invites a critical departure from such paradigms. While it is tempting to frame silkworms as alienated workers or unpaid outsourced labourers – contradicting Daoist advocacy of harmony and spontaneity and aligning with Marxist critiques of alienation – Liang's art complicates these binaries. By situating his work within the broader sericultural industry, I argue that silkworms' roles resist neat categorisation within human-defined notions of labour.

As delineated in preceding chapters, silkworm cultivation inherently exploits the silk-secreting capacities of this domesticated species, whose survival and reproduction are wholly dependent on human care. While silkworm husbandry appears modern under the global capitalist market – with large-scale production driven by crossbreeding and biotechnological advancements – it also retains its 'pre-modern' aspects. Silkworm-mulberry alliances demand intense, hands-on care from farmers, offering resistance to capitalist alienation, although as the very notion of a 'natural' relationship between silk and silkworm remains questionable. This raises questions not only about contemporary contexts but also about silkworm cultivation across history. Even in ancient times, silkworm cultivation as a livelihood was not necessarily more ideal or less exploitative and may not have aligned seamlessly with Daoist cosmology – a worthy line of inquiry when reconsidering human-silkworm relations diachronically. Thus, before entering the contemporary art arena, we must ask: Should silkworms already be theorised as labourers in these agricultural contexts? In Chapter 3, I experimented with framing silkworms as shamans in art rituals, emphasising their healing actions of covering and wrapping rather than reducing them to symbolic messengers or embodiments of social critique. However, recognising their generative and reproductive capacities as co-workers, shamans, or even co-authors to human artists raises a critical tension. Does such recognition risk projecting human divisions of labour, capitalist ideologies, and hegemonic aesthetic discourses onto nonhuman creativity, growth, and productivity (cf. Blanchette 2020)? These questions remain central for navigating how Liang's silkworm art inventively carved out zones for moral rumination.

As an anthropologist, my engagement with these pending perplexities is not to reason with abstract overloaded concepts but to interweave my thinking into how I design my research methods, select fieldwork materials, and the final narrative in this ethnography. Some anthropological approaches towards art have already placed more emphasis on the functional aspect and the generated social relations in artworks beyond their aesthetic and iconographic traits (e.g. Morphy 2013). The sensitivity to artistic skills (Gell 1998), to the social lives and orders of art objects (Miller 2015), to the perceptive flow of the artist, and to the surrounding material flow (Ingold 2013) has driven this book's exploration into the ritualistic art processes throughout Liang's and silkworms' shared life experiences. This sensitivity to flows aligns serendipitously with Daoist ethos of '*ziran*', where 'flow' is vaguely but omni-presently emphasised as the ultimate form of myriads of lives and relationships. As a method, my ethnography follows multispecies lives, energies, intentions, and forces that compose the material art objects, which are not discrete things, but circulate along with the worlds that they participate to compose in the vicinity. As a genre, this ethnography claims a 'parallel' programme of art criticism, experimenting non-alienated, grounded modes of sensing, interpreting and even presenting and preserving multispecies sociality in art practice.

As I wandered through the empty exhibition room at the Power Station of Art, I noticed the floor in the corner littered with 'Babies' – white cocoons wrapped in red silk swaddling clothes – exhibited as an isolated installation. A striking sign in front read 'No Touching' (Figure 6.1). Later, I asked Liang: wasn't it somewhat regrettable that viewers had encountered so many of your works without experiencing the nuanced tactile differences between flat silk and familiar silk products, or the sensation of caressing the body of

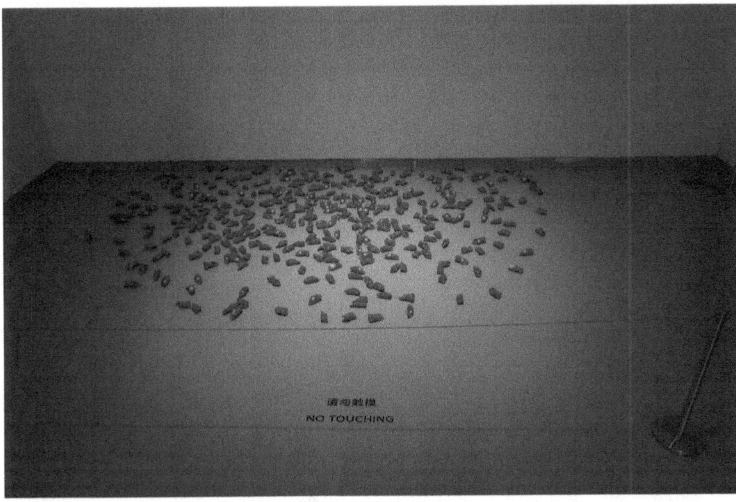

Figure 6.1 A 'No Touching' sign by 'Babies' in the gallery. Photo by author

silkworms? His response has since repeatedly lingered back to my mind: 'Regret is another space for creation and imagination'. If a unique theoretical-ethical programme of '*ziran*' emerges from Liang's silkworm art, 'regret' might well be its defining entry point. Consider 'Broken Landscape', born from the unprecedented rainy spring of 2005, when excessive silkworm urine devastated the spotless flat silk in Liang's studio. Or 'Self-roped', where beneath the seemingly harmonious human-silkworm symbiosis lies a silent, numbing endurance shared by both parties. The tender adoption cards, wishing silkworm 'babies' a long life and a future visit to China with their adopters, stand in stark contrast to the silkworms' ephemeral existence of just eight weeks. When interspecies harmony seems constantly unstable and transient, the resilient aliveness, prominent in Liang and the silkworms' artistic affinity, reveals itself through an optimism that refuses to be constrained by regret. Confronting inadequacies, imperfections, misses, and mistakes, such aliveness embodies not only the wisdom of accepting regret but the courage to take forward-moving actions without flinching, keeping the flows alive.

References

Blanchette, A., 2020. *Porkopolis: American Animality, Standardized Life, and the Factory Farm*. Durham: Duke University Press.

Dumit, J., 2014. Writing the Implosion: Teaching the World One Thing at a Time. *Cultural Anthropology*, 29 (2), 344–362.

Durkheim, E., 1912. *The Elementary Forms of the Religious Life*. New York: Free Press.

Geertz, C., 1973. *The Interpretation of Cultures*. New York: Basic Books.
Gell, A., 1998. *Art and Agency*. Oxford: Clarendon Press.
Gladston, P., 2011. A Conversation with Liang Shaoji. *In*: *Contemporary Art in Shanghai: Conversations with Seven Chinese Artists*. Hong Kong: Blue Kingfisher.
Gluckman, M., 1965. *Politics, Law and Ritual in Tribal Societies*. Chicago: Aldine.
Ingold, T., 2013. *Making: Anthropology, Archaeology, Art and Architecture*. London: Routledge.
La Berge, L.C., 2019. *Wages Against Artwork: Decommodified Labor and the Claims of Socially Engaged Art*. Durham: Duke University Press.
Leach, E., 1973. Levels of Communication and Problems of Taboo in the Appreciation of Primitive Art. *In*: A. Forge, ed. *Primitive Art & Society*. Cambridge: Cambridge University Press.
Miller, D., 2015. Materiality: An Introduction. In: D. Miller, ed. Durham: Duke University Press, 1–50.
Morphy, H., 2013. The Anthropology of Art. *In*: T. Ingold, ed. *Companion Encyclopedia of Anthropology*. London: Routledge, 648–685.
Mullin, M.H., 1999. Mirrors and Windows: Sociocultural Studies of Human-Animal Relationships. *Annual Review of Anthropology*, 28, 201–224.
Rappaport, R., 1968. *Pigs for the Ancestors: Ritual in the Ecology of a New Guinea People*. New Haven: Yale University Press.
Rapport, N. and Overing, J., 2000. *Social and Cultural Anthropology: The Key Concepts*. London: Routledge.
Turner, V., 1969. *The Ritual Process*. Suffolk: Penguin Books.
Turner, V., 1974. *Dramas, Fields and Metaphors*. Ithaca: Cornell University Press.
van Gennep, A., 1909. *The Rites of Passage*. Chicago: Chicago University Press.
von Drathen, D., 2011. Liang Shaoji: Of the Immaterial and Its Corporeality. *In*: M. Bloemheuvel and T. van Kooten, eds. *Windflower: Perceptions of Nature*. Rotterdam: NAi Publishers.
Xia, K., 2015. The Silk Shadow of Chora-topia [online]. *ShanghART Gallery*. Available from: https://www.shanghartgallery.com/galleryarchive/texts/id/8092.
Zulueta, C.C., 2016. Two Artists, Two Ecologies, and a Shared Empathy Towards Non-Human Animals' Agencies: Yanagi Yukinori and Liang Shaoji. *Journal of Contemporary Chinese Art*, 3 (3), 377–387.

Index

FSC
www.fsc.org
MIX
Papier | Fördert
gute Waldnutzung
FSC® C083411

Kolibri 360 GmbH
Ferdinand-Jühlke-Straße 7
99095 Erfurt, Deutschland
produktsicherheit@kolibri360.de